W9-ANH-684

Introduction to Theoretical Organic Chemistry and Molecular Modeling

Introduction to Theoretical Organic Chemistry and Molecular Modeling

William B. Smith

NEW YORK • CHICHESTER • WEINHEIM • BRISBANE • SINGAPORE • TORONTO

This book is printed on acid-free paper.

Originally published as ISBN 1-56081-937-5.

Published simultaneously in Canada.

Library of Congress Cataloging-in-Publication Data:
Smith, William B. (William Burton), 1927-
Introduction to theoretical organic chemistry and molecular modeling / by William B. Smith.
p. cm.
Includes bibliographical references.
ISBN 0-471-18643-0 (acid-free paper)
1. Chemistry, Physical organic. I. Title.
QD476.S567 1996
541.1'2--dc20 95-42883
CIP

Printed in the United States of America.

10 9 8 7 6 5 4 3 2

This volume is dedicated to

PAUL D. BARTLETT,

who taught many of us how to do
physical organic chemistry,

and to my wife,

MARIAN,

who made it possible for me to practice
the discipline.

"... chemistry is a tiny perturbation on the face of nature ..."

E.B. Wilson
The Robert A. Welch Foundation
Conference on Theoretical Chemistry
November 1972, Houston, Texas

"In every branch of physical knowledge there is nothing stable and certain but facts. Theories are as variable as the opinions that gave them birth. They are the meteors of the intellectual world, rarely productive of good, and more often hurtful to the intellectual progress of mankind."

Alexander Von Humboldt, 1796

"Progress of this kind (i.e., *developments in computational organic chemistry*) must, however, be paid for, and the price to the organic chemist is a proper understanding of quantum organic chemistry.
... Any organic chemist who lacks a proper understanding of current developments in MO theory is already heavily handicapped, and unless he takes steps to counter this, he will in a few years find himself left high and dry by the tide of theoretical organic chemistry ... ."

Michael J. S. Dewar, 1969

Preface

To learn the substance of all of modern organic chemistry is a daunting undertaking for the student. Whether by disinclination or a simple lack of time, most students of organic chemistry do not take enough courses in mathematics and physics to readily undertake courses in quantum mechanics. Yet one has only to open the covers of the most recent issue of the *Journal of the American Chemical Society* or the *Journal of Organic Chemistry* to realize that a large number of papers involving computational organic chemistry are now appearing. Molecular modeling in its various forms is being applied daily to problems of reaction mechanisms and synthesis. Each year vendors put forth a wealth of new computer programs for the organic chemist to use. For many, these programs remain in the realm of "black boxes." The manual may tell you how to do a calculation, but it usually will not tell you when you have made a mistake or when the calculation is meaningful. Sometimes it is the old adage "Garbage in, garbage out." One needs to understand what is going on in the black box to assure that the input poses a valid question and that the answer is a valid answer.

The development of quantum mechanics during the early decades of this century led to the detailed insight of the nature of the chemical bond, molecular structure, and chemical reactions which we enjoy today. Freshman college testbooks incorporate atomic orbital pictures, utilize the aufbau principle, and present verbalizations of the covalent bond derived from quantum theory. This picture is extended in the sophomore organic chemistry course. After a course in physical chemistry has been mastered, one should be prepared for the leap into theoretical organic chemistry at a useful level.

This text is the outcome of an earlier version dealing with quantum mechanics at a simple level. The early text was well received but has been in need of an overhaul for a number of years. The current text is addressed to organic chemistry students and extends the subject to molecular modeling via molecular mechanics and semiempirical and *ab initio* quantum mechanics, with a brief section on the rapidly developing field of density functional theory. Criticism was offered by some early reviewers of the book that too much time is spent with the Hückel molecular orbital (HMO) theory. However, many of the concepts and techniques of HMO theory are carried forward to more advanced techniques. Perturbation molecular orbital (PMO) and frontier molecular orbital (FMO) theories are derived from HMO theory and offer several advantages in understanding pericyclic and related reactions. The extended HMO was among the earliest approaches to handling saturated systems and still has applications today at the semiempirical level. Current semiempirical programs follow from this early work, which also provides a background for *ab initio* and density functional methods. The intent of the text is to lift the lid on the "black box" sufficiently to allow organic chemists to exploit the power of these techniques in a meaningful way.

The author wishes to express his appreciation to the large number of people who provided help toward the preparation of this volume. Some of this came in the form of gift software, and many contributed technical support on the use of various programs. Among these may be listed Kevin Gilbert (Serena Software), Doug Fox (Gaussian, Inc.), J.J.P. Stewart (MOPAC), Mathsoft (Mathcad), Trinity Software (HMO), and CAChe Scientific (CAChe). I would also like to thank editor Barbara Goldman of VCH who came to see that a book written for people who were already experts was not what the author intended.

William B. Smith
Fort Worth, TX

Contents

CHAPTER

1

The Hydrogen Atom and the Hydrogen Molecule

1.1 Introduction

The Bohr model of the hydrogen atom, introduced in 1913, represented a real break with the concepts of nineteenth-century physics because it stated that the electron moving about the nucleus was constrained to move in orbits in which the angular momentum was an integral multiple of $h/2\pi$. The h here is the ubiquitous Planck's constant, which had first appeared a few years earlier as a necessary proportionality constant in the equations accounting for black body radiation and the photoelectric effect. When accompanied by the assumption that energy could only be absorbed or emitted by the atom during transitions between these allowed energy states, the quantization of angular momentum led immediately to a highly satisfactory calculation of the lines of the hydrogen spectrum.

The success of the Bohr theory as applied to hydrogen was so immediate and complete that many assumed the final link in atomic structure to have been achieved. Unfortunately, attempts to apply the model to other atoms quickly revealed shortcomings that required more than a decade of work to correct.

Subsequent to the suggestion of de Broglie (1924) that high-energy particles might have both wave and particle characteristics, Schrödinger combined a set of quantum conditions with the wave equations of classical mechanics to produce a new description of matter.

1.2 The Schrödinger Equation

In classical particle physics it is often desirable to express the equation of motion of a particle or a group of particles by a function (the Hamiltonian) which embodies both the momenta and the coordinates of the system. For time-independent systems in classical motion

$$H = KE \text{ (Kinetic energy)} + V \text{(potential energy)}$$

The Hamiltonian here is simply the total energy of the system.

Consider a particle moving in one way along the x axis. Then,

$$KE = \frac{1}{2} mv^2 = \frac{p_x^2}{2m} \tag{1.1}$$

where p_x is the momentum along the axis. The total energy along the axis is

$$\frac{p_x^2}{2m} + V = E \tag{1.2}$$

The equation above can be converted to a form suitable for use in quantum mechanics by replacing the momentum with the operator $(h/2\pi i)(d/dx)$ (i is $\sqrt{-1}$) which gives us

$$-\frac{h^2}{8\pi^2 m} \frac{d^2}{dx^2} + V = E \tag{1.3}$$

The problem with Eq. 1.3 in this form is that it tells us to perform an operation on something but not what that something shall be.

Mathematical instructions to perform some task are often conveyed in the form of operators. In the expression $3 \times a$, the symbol $\times$ is an operator that tells us to multiply a by 3. Similarly, in the expression

$$\frac{d}{dx} e^{ax} = ae^{ax} \tag{1.4}$$

the (d/dx) symbol tells us to perform the differentiation.

The special form adopted in Eq. 1.4 is an example of a more general type of problem referred to as an eigenvalue problem. In eigenvalue problems, some function, here the eigenfunction e^{ax}, undergoes an operation that results in the formation of the eigenfunction multiplied by a constant. The constant, a, in Eq. 1.4, is referred to as the eigenvalue.

The Schrödinger equation is an eigenvalue problem. The left-hand side of Eq. 1.3 is an operator that is to act on a wave function ψ. When applied to the incomplete statement of Eq. 1.3 and rearranged, Eq. 1.5 results:

$$-\frac{h^2}{8\pi^2 m}\frac{d^2\psi}{dx^2} + V\psi = E\psi$$

$$\frac{d^2\psi}{dx^2} + \frac{8\pi^2 m}{h^2}(E - V)\psi = 0 \tag{1.5}$$

This equation applies to a particle moving in one dimension, and is a simple form of the Schrödinger equation. In three dimensions we must resort to partial differentiation, and Eq. 1.5 becomes

$$\nabla^2\psi + \frac{8\pi^2 m}{h^2}(E - V)\psi = 0 \tag{1.6}$$

where ∇^2 is the Laplacian operator

$$\frac{\partial^2}{\partial x^2} + \frac{\partial^2}{\partial y^2} + \frac{\partial^2}{\partial z^2}$$

In a condensed form, the Schrödinger equation is written as follows:

$$\mathbf{H}\psi = E\psi \tag{1.7}$$

In keeping with all differential equations, such as Eqs. 1.6 and 1.7, there are a large number of available solutions, but the physical nature of the problem imposes certain limitations. The physical interpretation of ψ is that of a probability distribution such that the probability of finding the electron at a given point in space is given by ψ^2. A more nearly correct way of expressing this is to say that the probability of finding the electron in a certain small volume element $d\tau$ ($dxdydz$ in Cartesian coordinates) is $\psi^2 d\tau$. Since the probability of finding the electron somewhere in space is 1, the value of the integral $\int \psi^2\, d\tau$ taken over all space must be 1. Wave functions that have this property are said to be normalized. For an unnormalized function, $\int \psi^2\, d\tau = N^2$ where N is a constant, the normalization factor. The normalized wave function in the latter case becomes ψ/N.

Since the probability of finding the electron at any point in space is real, the Ψ function must be finite. Similarly, since no sudden changes in probability are expected, the function must be continuous, and the function must be single valued, as there is only one probability of finding the electron at a given point in space.

One possible solution to Eq. 1.6 is given by

$$\psi = e^{-kr}$$

This solution corresponds to the hydrogen 1*s* state, the lowest energy state available to an electron in the hydrogen atom. Other solutions describe the 2*s*, 2*p*, etc., states (for a general expression for these, see Chapter 7).

Application of the Schrödinger equation to other atoms in the periodic table requires approximate methods, since one is faced with the insoluble classical three-body problem. However, very good wave functions are now available taht account well for most, if not all, atomic spectra. These approximate wave functions not only give the the electronic energy associated with each state, but also describe that region of space in which the probability of finding the electron is high. This last result gives us the atomic orbital (AO) pictures that decorate so many texts. As drawn, these orbitals are usually shown as that region of space enclosing 90% of the electron density. Wave functions for atoms above hydrogen in the periodic table are usually generated by altering the constants in the hydrogen wave functions and so have the same general form.

1.3 The Hydrogen Molecule

In attempting to find wave functions that will satisfactorily account for the energy states found in the hydrogen molecule, one is immediately faced with an insoluble problem regarding the equations of motion for four bodies—two protons and two electrons. A variety of approximate methods have been explored, but we will follow only the line that leads to our goal.

The electrons in molecules can be pictured as occupying orbitals in much the same sense as those associated with atoms. Each such electron is described by a Ψ function such that Ψ^2 gives the probability distribution. That volume of space in which the probability density is high is pictured as the molecular orbital in much the same way as in atoms. For each Ψ there is an associated energy that approximates to the negative of the ionization energy for the electron. Electrons are assigned to molecular orbitals (MOs) in much the same way as to AOs, and Hund's rule regarding electron pairing applies, i.e., electrons will only pair up when all orbitals of the same energy are at least singly occupied.

The next problem is how to approximate the wave functions for the MOs. The approximation used here is that of a *linear combination of atomic orbitals* (LCAO), which for hydrogen assumes the form

$$\Psi = c_1\psi_1 + c_2\psi_2$$

where ψ_1 and ψ_2 are hydrogen 1*s* wave functions. This form follows from the expectation that in the region of one of the atoms the electron will be strongly influenced by that atom and will be described by a wave function similar to the atomic wave function. The coefficients in the LCAO must be adjusted to minimize the total energy of the molecule, and for this purpose one makes use of the *method of variations.*

The energy term of the Schrödinger equation can be extracted by multiplying both sides by Ψ and integrating over all coordinates involved:

$$\Psi \mathbf{H} \Psi = \Psi E \Psi$$

$$E = \frac{\int \Psi \mathbf{H} \Psi \, d\tau}{\int \Psi^2 \, d\tau} \tag{1.8}$$

There are several reasons for this operation rather than simply settling for the more obvious device of dividing both sides by Ψ. The function $H\Psi/\Psi$ frequently has no satisfactory minimum property. Furthermore, Ψ can have a zero value making the previous expression go to infinity.

Substitution of the LCAO gives

$$E = \frac{\int (c_1\psi_1 + c_2\psi_2) H (c_1\psi_1 + c_2\psi_2) \, d\tau}{\int (c_1\psi_1 + c_2\psi_2)^2 \, d\tau}$$

$$= \frac{c_1^2 \int \psi_1 H \psi_1 \, d\tau + 2c_1c_2 \int \psi_1 H \psi_2 \, d\tau + c_2^2 \int \psi_2 H \psi_2 \, d\tau}{c_1^2 \int \psi_1^2 \, d\tau + 2c_1c_2 \int \psi_1\psi_2 \, d\tau + c_2^2 \int \psi_2^2 \, d\tau} \tag{1.9}$$

This relation can be compressed by the following type of shorthand:

$$H_{11} = \int \psi_1 H \psi_1 \, d\tau$$

The integrals H_{11} and H_{22} are called *Coulomb integrals* and, of course, are equal to each other in this problem. In the limit where the nuclei are far apart, the Coulomb integral reduces to the expression for the electron energy in atomic hydrogen. However, at closer distances the integral represents the energy of an electron moving in an orbital about one atom in the presence of the second. Hence,

$$H_{12} = \int \psi_1 H \psi_2 \, d\tau = \int \psi_2 H \psi_1 \, d\tau$$

This equality is proven in a number of texts on quantum mechanics.

The integral H_{12} expresses the energy of the electron moving under the influence of both nuclei and is known as the *bond* or *resonance* integral. The value of H_{12} is negative, as is that for H_{11}.

$$S_{11} = \int \psi_1 \psi_1 \, d\tau$$

This integral appears to have no specific name. In the case of the hydrogen atom, this function is the probability of finding the electron somewhere in space. The wave functions are selected in such a way that the value of the integral is one. Such wave functions are said to be *normalized*. We will work only with normalized wave functions hereafter.

$$S_{12} = \int \psi_1 \psi_2 \, d\tau$$

The integral S_{12} is a measure of the overlap of the hydrogen atomic 1*s* orbitals in forming the new molecular orbitals. When the AOs are far apart, the value of S_{12} is zero. As the orbitals approach each other, the value of S_{12} increases. At the equilibrium nuclear distance for the hydrogen molecule, S_{12} is 0.59. The integral may be zero under other conditions too, i.e., when considering the interaction of an *s* orbital and a *p* orbital in the same atom. When the value of S_{12} is zero, the wave functions are said to be *orthogonal*.

Placing these definitions in Eq. 1.9 leads to

$$E = \frac{C_1^2 H_{11} + 2C_1C_2H_{12} + C_2^2 H_{22}}{C_1^2 + 2C_1C_2S_{12} + C_2^2} \tag{1.10}$$

In order to find those C values that will minimize the energy, one now differentiates E with respect to C_1 and C_2:

$$\frac{\partial E}{\partial C_1} = \frac{2C_1H_{11} + 2C_2H_{12}}{(C_1^2 + 2C_1C_2S_{12} + C_2^2)} - \frac{(C_1^2H_{11} + 2C_1C_2H_{12} + C_2^2H_{22})(2C_1 + 2C_2S_{12})}{(C_1^2 + 2C_1C_2S_{12} + C_2^2)^2}$$

$$= \frac{2C_1H_{11} + 2C_2H_{12} - E(2C_1 + 2C_2S_{22})}{(C_1^2 + 2C_1C_2S_{12} + C_2^2)} = 0$$

Therefore

$$C_1(H_{11} - E) + C_2(H_{12} - ES_{12}) = 0 \tag{1.11}$$

and correspondingly from the $(\partial E/\partial C_2)$,

$$C_1(H_{12} - ES_{12}) + C_2(H_{22} - E) = 0 \tag{1.12}$$

Equations 1.11 and 1.12 are called the secular equations. For a nontrival solution for these linear homogeneous equations (i.e., for a solution other than $C_1 - C_2 = 0$), the secular determinant must be zero.

$$\begin{vmatrix} H_{11} - E & H_{12} - ES_{12} \\ H_{12} - ES_{12} & H_{22} - E \end{vmatrix} = 0 \tag{1.13}$$

This determinant is of the form

$$\begin{vmatrix} a_{11} & a_{12} \\ a_{21} & a_{22} \end{vmatrix} = 0$$

and the solution can be written explicitly as $a_{11} \times a_{22} - a_{21} \times a_{12} = 0$. It is left as an exercise to convince oneself that expansion of the determinant and solution of the resulting quadratic in E gives

$$E = \frac{H_{11} \pm H_{12}}{1 \pm S_{12}} \tag{1.14}$$

Theoreticians, like organic chemists, sometimes become bemused by nomenclature. The square arrays above are sometimes referred to as determinants (often secular determinants), and that term will be used here. Textbooks on linear algebra, determinants, and matrices frequently state that a determinant is a number, whereas others would say that the quadratic equation that results from expanding Eq. 1.13 is the determinant. The subject of determinants, matrices, and methods to work with them is discussed at greater length in Appendix A.

The solutions of the secular determinant for the hydrogen molecule thus give two energy states, and both electrons will occupy the lower (or ground state) level.

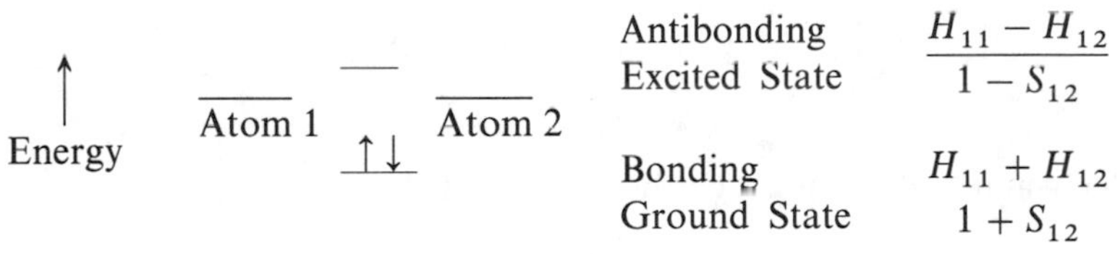

Energy ↑	Atom 1		Atom 2		
		—		Antibonding Excited State	$\frac{H_{11} - H_{12}}{1 - S_{12}}$
		↑↓		Bonding Ground State	$\frac{H_{11} + H_{12}}{1 + S_{12}}$

Because of the negative values associated with H_{11} and H_{12}, the form of Eq. 1.14 with the positive signs corresponds to the lower state.

Calculations of the total energy of the molecule will require evaluation of the integrals defined above, which, in turn, implies a knowledge of the Hamiltonian expressing all the attractive and repulsive interactions between the two electrons and the two protons. Happily, in carbocyclic systems it is usual to evaluate these integrals empirically.

For the hydrogen molecule ion (H_2^+), the problem and its solutions are formally as above. However, the equilibrium bond length will not be the same as for the neutral molecule, and each of the integrals will assume a different value. This observation is made here to forewarn us that the assumption that these integrals will be the same for carbocyclic neutral and charged systems as for heteroatom systems will prove to be an oversimplification.

Using the above values for E and the secular equations, it is possible now to solve for the coefficients. This will be demonstrated more explicitly in the next chapter and in Appendix A. Substitution into the LCAO expression gives the two wave functions—one for each energy level:

$$\Psi_1 = \left(\frac{1}{\sqrt{2(1 - S_{12})}}\right)(\psi_1 + \psi_2)$$

$$\Psi_2 = \left(\frac{1}{\sqrt{2(1 - S_{12})}}\right)(\psi_1 - \psi_2)$$

Cross sections of these two wave functions may be sketched from the individual hydrogen atomic wave functions (showing only the radial portion of the wave function):

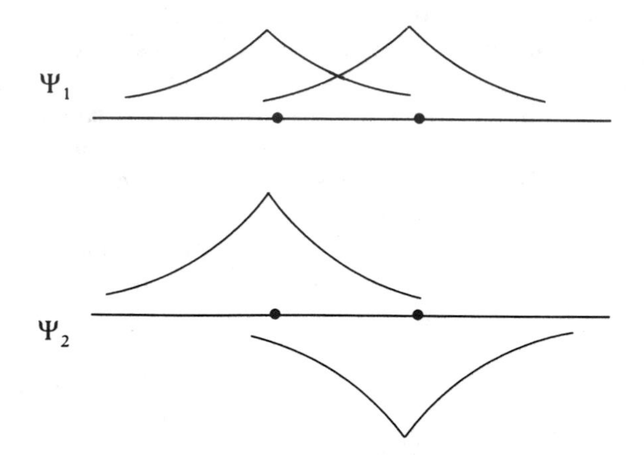

The electron density is given by the squares of the Ψ values:

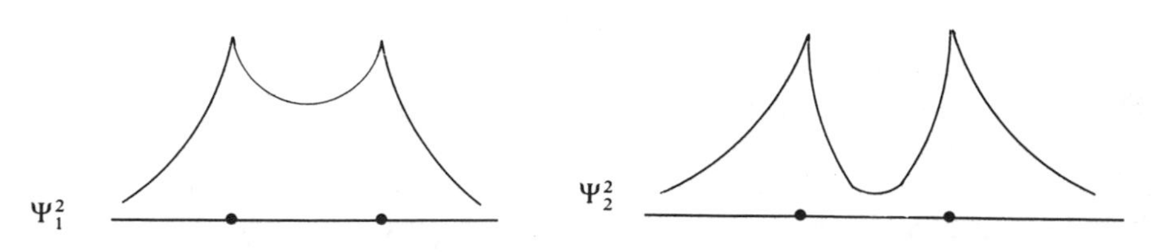

As can be seen, the squaring of the wave functions leads to increased electron density between the two nuclei for the filled lower energy level (here denoted by Ψ_1^2). The energy of the two electrons moving under the influence of both protons results in bond formation and a lowering of the energy compared to the sum for the two isolated atoms. For Ψ_2 at a point midway between the two atoms, the probability Ψ_2^2 is zero. This point is called a node. MOs that have a nodal point (or a plane in three dimensions) in between the nuclei are called *antibonding* MOs.

It may occur to some that no accounting has been made of the vibrations of the protons about which the electrons move. These atomic nuclei do move, of course, and the energy of the molecule and the motions of the electrons should be affected by the nuclear motions. However, the mass of a proton is three orders of magnitude larger than that of the electron. The latter move with so much greater velocity that for practical purposes the nuclear motion is nil. The statement of this as a principle was first made by Born and Oppenheimer, and we will assume in all that follows that nuclear motion can be ignored. The hydrogen molecule is a worst-case scenario, and even here the error introduced by the Born-Oppenheimer approximation is trivial.

Problems

1. If ψ_1 and ψ_2 are each normalized and orthogonal (i.e., orthonormal functions), show that the LCAO wave functions below are also orthogonal.

$$\Psi_1 = \frac{1}{\sqrt{2}}\psi_1 + \frac{1}{\sqrt{2}}\psi_2$$

$$\Psi_2 = \frac{1}{\sqrt{2}}\psi_1 - \frac{1}{\sqrt{2}}\psi_2$$

2. In the classical Hooke's law problem of a vibrating weight on a spring, the potential energy is given by $\frac{1}{2}kx^2$ where k is the force constant reflecting the strength of the spring and x is the displacement from the equilibrium position. The classical vibration frequency ν is given by the expression $(1/2\pi)\sqrt{k/m}$. For a comparable molecular vibration, the one-dimensional Hooke's law expression becomes

$$\frac{\partial^2\psi}{\partial x^2} + \frac{8\pi^2 m}{h^2}\left(E - \frac{1}{2}kx^2\right)\psi = 0$$

where the ground state wave function can be shown to be of the form $\psi_0 = e^{-\alpha x^2}$. You are to find α in terms of m and k; then find E_0 in the same terms.

CHAPTER

2

Hückel Molecular Orbital Theory

2.1 The Concept of Directed Valence and Hybridization

In Chapter 1 the MOs of the hydrogen molecule were formed by overlapping the 1*s* orbitals of each hydrogen atom. If we carry this line of thinking forward to a more complex molecule such as water, the expectation would be that the oxygen (electronic configuration $1s^2 2s^2 2p_x^2 2p_y^1 2p_z^1$) would overlap its two *p* orbitals with two hydrogen 1*s* orbitals to form two covalent bonds. The predicted H—O—H bond angle would be 90°, in keeping with the orientation of the *p* orbitals. Such a guess would not be too far in error, as the experimental value for water is 104°. We might even rationalize the difference as being due to electronic repulsions between the electrons making up the covalent bonds. A similar argument might be used in combining nitrogen ($1s^2 2s^2 2p_x^1 2p_y^1 2p_z^1$) with three hydrogens to form ammonia, though here the predicted 90° angles are even farther afield from the experimental value of 107°.

This line of thinking fails completely when one considers the bonding of carbon ($1s^2 2s^2 2p_x^1 2p_y^1$). Carbon is tetravalent and displays geometries in its varied compounds from linear (C coordination number 2) to planar trigonal (coordination number 3) and tetrahedral (coordination number 4). This problem was first attacked in the late 1920s by Linus Pauling and led to the concept of orbital hybridization.

The mathematical descrptions of atomic *s*, *p*, *d* ... etc. orbitals are each solutions to a differential equation. Freshman and sophomore chemistry textbooks (and, one supposes, high school texts as well) are filled with artistic representations of these orbitals. One quality of such solutions is that linear combinations of these are themselves solutions to the original equation. So we might write

$$\Psi_1 = a_1\psi_{2s} + b_1\psi_{2px} + c_1\psi_{2py} + d_1\psi_{2pz}$$

$$\Psi_2 = a_2\psi_{2s} + b_2\psi_{2px} + c_2\psi_{2py} + d_2\psi_{2pz}$$

$$\Psi_3 = a_3\psi_{2s} + b_3\psi_{2px} + c_3\psi_{2py} + d_3\psi_{2pz}$$

$$\Psi_4 = a_4\psi_{2s} + b_4\psi_{2px} + c_4\psi_{2py} + d_4\psi_{2pz}$$

where the small ψ values are the atomic orbitals for carbon. These are then solved for the coefficients (a_i, b_i, c_i, and d_i) with the imposition of the condition that the hybrid orbitals must be normalized and orthogonal. The result is the familiar arrangement of four orbitals with lobes pointing to the vertices of a regular tetrahedron (tetrahedral hybridization). If only the $2s$, $2p_x$, and $2p_y$ atomic orbitals are used, the result is three new planar sp^2 hybridized orbitals with 120° angles (trigonal) and an orthogonol unhydridized $2p_z$ orbital. If only the $2s$ and $2p_x$ orbitals are hydridized, the new hybrids are colinear linear orbitals with $2p_y$ and $2p_z$ orbitals not involved in the hydridization process. To a good approximation, the geometries of methane, ethylene, and acetylene can be described as involving carbon in these different states of hybridization. A representation of the three types of hybrid carbon orbitals is given in Figure 2.1.

In the material immediately to follow, interest will be centered on sp^2 trigonal carbons. The sigma bonds of ethylene are formed by the overlap of trigonal sp^2 bonds between the two carbons and the four hydrogens. Our interest will center on what happens to the $2p_z$ orbitals which were not involved in the hybridization process.

Linus Pauling in his classic text *The Nature of the Chemical Bond* (Cornell University Press, 1948) made the point that "of two orbitals in an atom the one which can overlap more with an orbital of another atom will form the stronger bond with that atom, and, moreover, the bond formed by a given orbital will tend to lie in that direction in which the orbital is concentrated." Thus, the concept of orbital hybridization offered an explanation of the geometry of carbon bonds and went far to explain the differences in geometry and chemical reactivity of carbon compounds. We will see in later chapters that he assumption of orbital hybridization is not required in modern all-electron computations. These programs use one of several approximations for the AOs and automatically correct the orbital mix to give a minimum energy electron configuration and geometry to the molecule. However, we will stick to the hybridization construct until we have laid the foundation for those calculations.

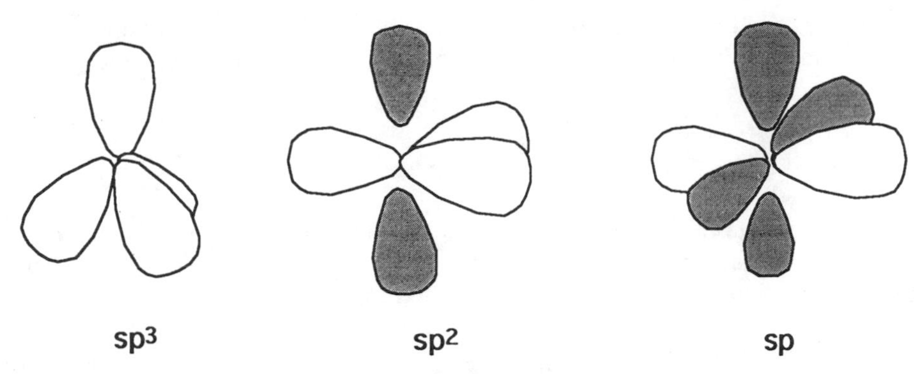

Figure 2.1. The various hybrid carbon orbitals are indicated as open orbitals, and the *p* orbitals not involved in the hybridization are shaded.

2.2 More on the Method of Variations

Before discussing the Hückel theory in detail, let us provide some background that expands upon the discussion initiated in Chapter 1. In many cases the LCAO approximation will serve as a basis to approximate the true wave functions for very complex situations, and we will have occasion to refer to it frequently. It goes without saying that one then must employ the method of variations to produce a set of coefficients for the basis functions that make up the LCAO.

For a general case we can write

$$\Psi = c_1\psi_1 + c_2\psi_2 + c_3\psi_3 + \cdots + c_n\psi_n = \sum_i c_i\psi_i \tag{2.1}$$

where the various ψ functions are AOs and the Ψ are MOs associated with energies $E_0, E_1, E_2 \ldots E_n$. If our wave function were an exact wave function, these energies would, of course, be exact energies derived from the Schrödinger equation. However, with approximate wave functions only approximate energies called expectation values (sometimes given as $\langle E \rangle$) can be found:

$$E = \frac{\int \Psi \mathbf{H} \Psi \, d\tau}{\int \Psi\Psi \, d\tau} \tag{2.2}$$

The closer the approximation for the wave functions, the nearer will the expectation values come to the true values.

It is necessary now to substitute the expression on the right side of Eq. 2.1 in place of Ψ in Eq. 2.2. As in Chapter 1, we must minimize the energy with respect to each of the coefficients by a series of n partial differentiations, setting each equal to zero, i.e., the coefficients are treated as the variables and

$$\frac{\partial E}{\partial c_1} = 0 \qquad \frac{\partial E}{\partial c_2} = 0 \ldots \frac{\partial E}{\partial c_i} = 0 \tag{2.3}$$

We won't write these processes out in detail, since they are but more complex forms of the earlier calculation for the hydrogen molecule. As there, the result is a series of simultaneous equations of the form

$$\begin{aligned}
&c_1(H_{11} - S_{11}E) + c_2(H_{12} - S_{12}E) + c_3(H_{13} - S_{13}E) + \cdots + c_n(H_{1n} - S_{1n}E) = 0 \\
&c_1(H_{21} - S_{21}E) + c_2(H_{22} - S_{22}E) + c_3(H_{23} - S_{23}E) + \cdots + c_n(H_{2n} - S_{2n}E) = 0 \\
&\vdots \\
&c_1(H_{n1} - S_{n1}E) + c_2(H_{n2} - S_{n2}E) + c_3(H_{n3} - S_{n3}E) + \cdots + c_n(H_{nn} - S_{nn}E) = 0
\end{aligned}$$

It will be stated without proof that a both necessary and sufficient condition for the above equations to be true is that the determinant of the coefficients must equal zero:

$$\begin{vmatrix}
H_{11} - S_{11}E \; H_{12} - S_{12}E \; H_{13} - S_{13}E \ldots H_{1n} - S_{n1}E \\
H_{21} - S_{21}E \; H_{22} - S_{22}E \; H_{23} - S_{23}E \ldots H_{2n} - S_{21}E \\
H_{11} - S_{11}E \; H_{12} - S_{12}E \; H_{13} - S_{13}E \ldots H_{3n} - S_{31}E \\
\vdots \\
H_{n1} - S_{n1}E \; H_{n2} - S_{n2}E \; H_{n3} - S_{n3}E \ldots H_{nn} - S_{nn}E
\end{vmatrix} = 0 \tag{2.4}$$

where the H_{ij} and S_{ij} terms are as previously defined.

Determinants above the second order cannot be dealt with by simple cross multiplication, as with the hydrogen molecule. Questions regarding the determination of the eigenvalues (energies) and eigenvectors (coefficients) are addressed in detail in Appendix A, and it will be assumed throughout the balance of this chapter that techniques exist for such determinations. For the sake of convenience and future use, the determinant above may be condensed to

$$\sum_{i=1}^{n} c_{ij}[H_{ij} - ES_{ij}] = 0 \qquad j = 1, 2, 3, \ldots, n \tag{2.5}$$

Since not all of the c_{ij} values can be zero, it follows that the value of the determinant must equal zero:

$$|H_{ij} - S_{ij}E| = 0 \tag{2.6}$$

2.3 The Hückel Theory

In 1931–32, E. Hückel published [*Z. Physik*, **70**, 204 (1931) and **76**, 628(1932)] the first applications of quantum mechanics to large organic molecules. In view of the state of the art then, it is understandable that some rather drastic approximations were required. Although it has often been criticized in more recent years, the fact remains that the Hückel method provided a number of extremely interesting results for the time. Since the early 1970s, professional theoretical organic chemists have employed increasingly more sophisticated methods, but for some purposes Hückel theory remains of value. It continues to serve as a worthwhile introduction to the more complex theories for the uninitiated.

Hückel's method applies to planar conjugated systems and depends immediately on the assumption that σ- and π-electrons in bonds act independently. Although this assumption seriously affects calculations of electron spin resonance phenomena, it seemingly does not greatly influence many calculations of interest.

The Coulomb integrals H_{ii} between sp^2 hybridized carbons are assumed to be equal and to approximate the energy of an electron in a carbon $2p$ orbital. Hereafter, the Coulomb integral will be designated as α and taken as having a negative value.

The resonance integral H_{ij}, called β hereafter, and the overlap integral S_{ij} should obviously be a function of the length of the bond involved. In the Hückel molecular orbital (HMO) method, it is usual to assume all β values are the same, although the variation of β with bond length can be incorporated if one wishes. Resonance integrals between nonadjacent carbons are set at zero. Moreover, in HMO calculations the obvious oversimplification is made that overlap integrals are zero. This greatly simplifies the mathematics and produce results not grossly different from those including reasonable values for S. In computer-based calculations, S values are often incorporated.

2.4 Ethylene

The problem of the π bond in ethylene is formally similar to the hydrogen molecule problem, in that two AOs (in this case carbon $2p$ orbitals) are used in forming two new MOs. Equation 2.6, modified for the Hückel assumptions and with definitions above, can now be written as

$$\begin{vmatrix} \alpha - E & \beta \\ \beta & \alpha - E \end{vmatrix} = 0 \tag{2.7}$$

Dividing both rows by β and setting x equal to $(\alpha - E)/\beta$ gives an even simpler form

$$\begin{vmatrix} x & 1 \\ 1 & x \end{vmatrix} = 0 \tag{2.8}$$

from which we derive

$$X^2 - 1 = 0$$

$$x = \pm 1 \quad \text{and} \quad E = \alpha \pm \beta \tag{2.9}$$

Again bearing in mind that α and β are negative entities, it is seen that $\alpha + \beta$ represents the ground state energy of the ethylene π system.

Level	*Energy*
——	$\alpha - \beta$
$\uparrow\downarrow$	$\alpha + \beta$

$$E_{\pi}^{\text{total}} = 2\alpha + 2\beta$$

From the secular equations

$$C_1(\alpha - E) + C_2\beta = 0$$

$$C_1\beta + C_2(\alpha - E) = 0$$

When $E = \alpha + \beta$ then

$$\frac{C_1}{C_2} = 1$$

and when the condition $\Sigma C_i^2 = 1$ is applied, it may be seen by inspection that $C_1 = C_2 = \sqrt{1/2}$. Similarly, for the root $\alpha - \beta$, $C_1 = -C_2 = 1/\sqrt{2}$. The two wave functions for the bonding and antibonding π electron states of ethylene are

$$\Psi_1 = \frac{1}{\sqrt{2}}\psi_1 + \frac{1}{\sqrt{2}}\psi_2$$

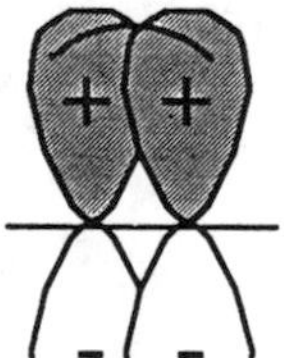

+ +
- -

$$\Psi_2 = \frac{1}{\sqrt{2}}\psi_1 - \frac{1}{\sqrt{2}}\psi_2$$

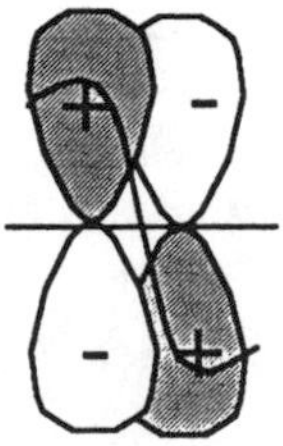

+ -
- +

As shown, the antibonding function has a node between the carbons—a typical feature of antibonding states.

2.5 Butadiene

In butadiene ($CH_2{=}CH{-}CH{=}CH_2$) the π-electrons occupy MOs made up of combinations of the four carbon $2p$ orbitals, i.e.,

$$\Psi_i = c_1\psi_1 + c_2\psi_2 + c_3\psi_3 + c_4\psi_4 \tag{2.10}$$

Applying the HMO postulates and the abbreviations, one can write as follows:

$$\begin{vmatrix} \alpha - E & \beta & 0 & 0 \\ \beta & \alpha - E & \beta & 0 \\ 0 & \beta & \alpha - E & \beta \\ 0 & 0 & \beta & \alpha - E \end{vmatrix} = \begin{vmatrix} x & 1 & 0 & 0 \\ 1 & x & 1 & 0 \\ 0 & 1 & x & 1 \\ 0 & 0 & 1 & x \end{vmatrix} = 0 \tag{2.11}$$

Consult Appendix A for the method of determining the eigenvalues (energies) and eigenvectors (coefficients) represented by the secular determinant Eq. 2.11. The fourth-order polynomial derived from Eq. 2.11 gives four eigenvalues: $x = \pm 0.62$ and ± 1.62. Thus, the values of E in ascending order of energy are as shown below. The four π electrons fill only orbitals 1 and 2.

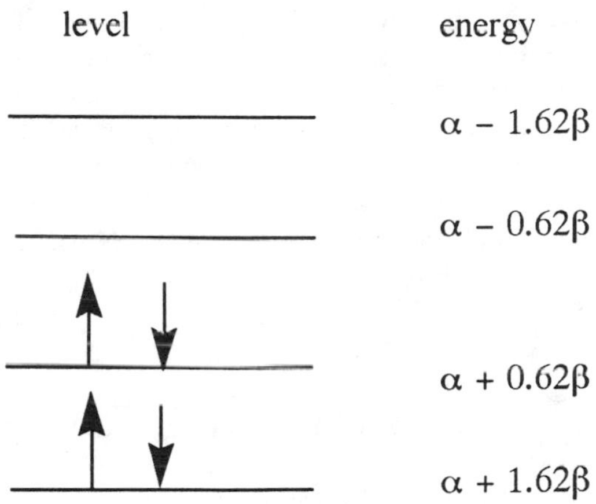

level	energy
———	$\alpha - 1.62\beta$
———	$\alpha - 0.62\beta$
↑↓	$\alpha + 0.62\beta$
↑↓	$\alpha + 1.62\beta$

$$\mathbf{E_{\pi}^{total} = 4\alpha + 4.48\beta}$$

In classical butadiene the π electrons are localized in bonds between C-1 and C-2 and between C-3 and C-4, respectively. Since no interaction occurs between the C-2 and C-3 $2p$ orbitals, the value of β goes to zero for the 2–3 π bond. The secular determinant is altered to

$$\begin{vmatrix} x & 1 & 0 & 0 \\ 1 & x & 0 & 0 \\ 0 & 0 & x & 1 \\ 0 & 0 & 1 & x \end{vmatrix} = 0$$

This determinant is in block diagonal form and factors directly to two 2×2 determinants identical to that for ethylene. The four roots then are $\alpha \pm \beta$ and $\alpha \pm \beta$. For this form of butadiene, $E_{\pi}^{\text{total}} = 4\alpha + 4\beta$, which is simply twice the energy of ethylene.

The energy difference between delocalized and classical (localized) butadiene is 0.48 β. This delocalization energy corresponds roughly to the concept of resonance energy as conceptualized by the organic chemist, and is measured by heats of combustion or hydrogenation. Since β cannot be readily evaluated directly, it is generally assigned some experimental magnitude. The above treatment applied to benzene gives a delocalization energy (DE) of 2β. Since the experimental resonance energy for benzene is 36 kcal/mol, a value of 18 kcal/mol is convenient for β and gives fairly satisfactory agreement between experimental and HMO values for the DE in a number of aromatic hydrocarbons. We must keep in mind that the Hückel approximations are now approximately 60 years old. Through the years the oversimplications inherent in this system have often been attacked. Professor Dewar has provided an argument that the above value is better represented by a value of 10 kcal/mol, though this does not give a calculated value in agreement with the experimental thermochemical value.[1] It is certainly true that the values of β vary with bond lengths and are not the same for both the classical double bond and those found in delocalized systems. These matters will be treated in more detail. The calculated DE for butadiene on the naive basis is 8.5 kcal/mol as compared to an experimental value of 3.5.

Additional information about the butadiene molecule depends on a knowledge of the complete set of MO wave functions. The coefficients may be determined as before, i.e., by substituting each root, one at a time, into the secular equations and solving simultaneously. Normalization must also be applied. A more systematic and generally applicable method is set forth in Appendix A. Since a number of computer programs are now available for HMO calculations, the matter will not be treated in greater detail here.[2]

The wave functions for butadiene, in ascending order of energy, are

$$\Psi_1 = 0.372\psi_1 + 0.602\psi_2 + 0.602\psi_3 + 0.372\psi_4$$
$$\Psi_2 = 0.602\psi_1 + 0.372\psi_2 - 0.372\psi_3 + 0.602\psi_4$$
$$\Psi_3 = 0.602\psi_1 - 0.372\psi_2 - 0.372\psi_3 + 0.602\psi_4$$
$$\Psi_4 = 0.372\psi_1 - 0.602\psi_2 + 0.602\psi_3 - 0.372\psi_4$$

A half-tone rendering of the resultant orbitals is given below. The highest occupied orbital (HOMO) and the lowest unoccupied orbital (LUMO) are indicated. The lobes that are darker correspond to positive values of the MO.

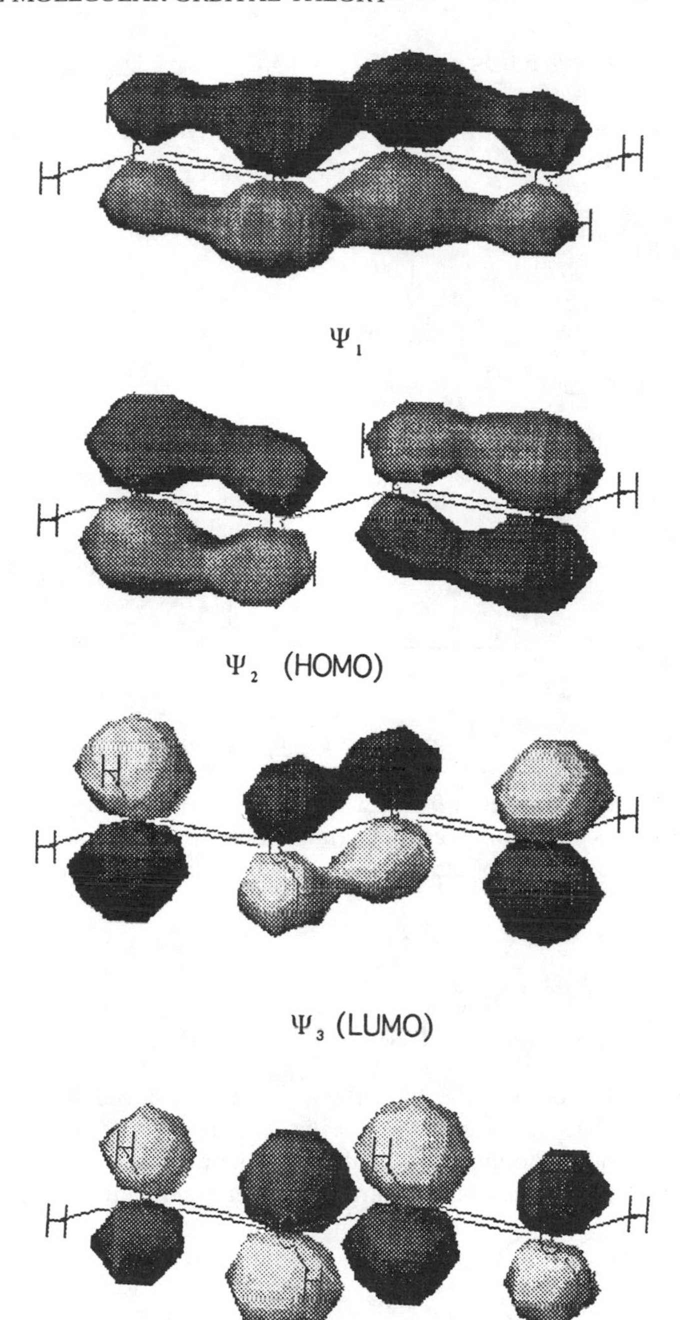

Ψ_4

2.6 Nonbonding MOs — Allyl

The three-carbon π system that is present in the allyl cation, free radical, and anion leads to the following secular determinant:

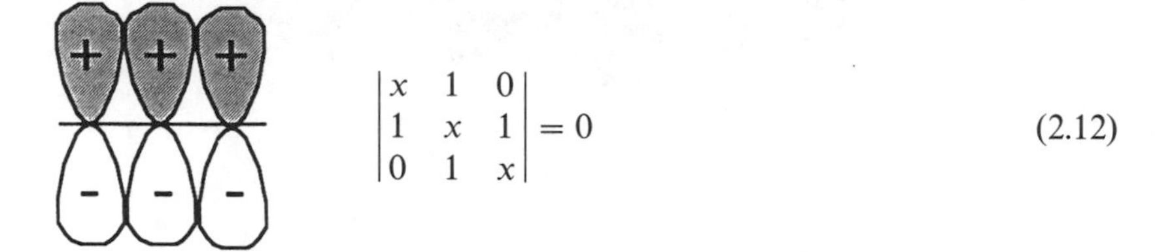

$$\begin{vmatrix} x & 1 & 0 \\ 1 & x & 1 \\ 0 & 1 & x \end{vmatrix} = 0 \qquad (2.12)$$

which leads in turn, to $x = -\sqrt{2}$ and $+\sqrt{2}$. The available energy levels are

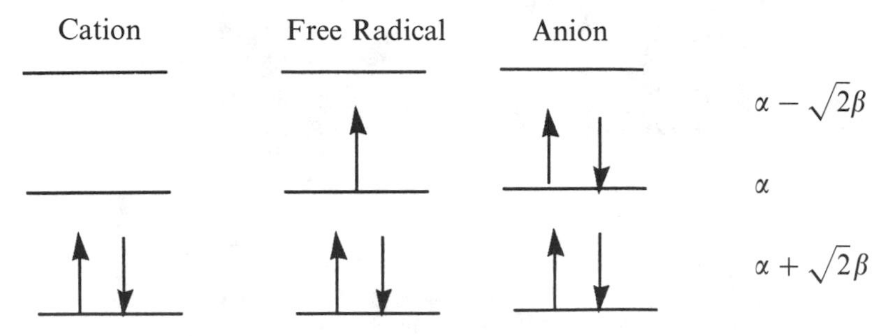

In the cation the first level is filled, while the radical contains these plus one electron in the second orbital, and the anion contains a full first level plus two electrons in the second orbital. The wave functions can be shown to be

$$\Psi_1 = 1/2\psi_1 + 1/\sqrt{2}\psi_2 + 1/2\psi_3$$

$$\Psi_2 = 1/\sqrt{2}\psi_1 - 1/\sqrt{2}\psi_3$$

$$\Psi_3 = 1/2\psi_1 - 1/\sqrt{2}\psi_2 + 1/2\psi_3$$

The electrons that occupy the Ψ_2 orbital are characterized by an energy α approximating that of an electron localized in a carbon $2p$ orbital. Such electrons contribute nothing to the stability of the molecule. A diagram of Ψ_2 shows a node at C-2. Such orbitals are called nonbonding molecular

(2.13)

orbitals (NBMOs), and will be discussed in considerable detail subsequently.

2.7 Degenerate Orbitals — Cyclopropenyl

The cyclopropenyl cation has been of interest because of the preditions of a large delocalization energy. The secular determinant and roots for the system are as follows:

$$\begin{vmatrix} x & 1 & 1 \\ 1 & x & 1 \\ 1 & 1 & x \end{vmatrix} = 0$$

$$X = +1, +1, -2$$

$$E = \alpha+2\beta, \ \alpha-\beta, \ \alpha-\beta$$

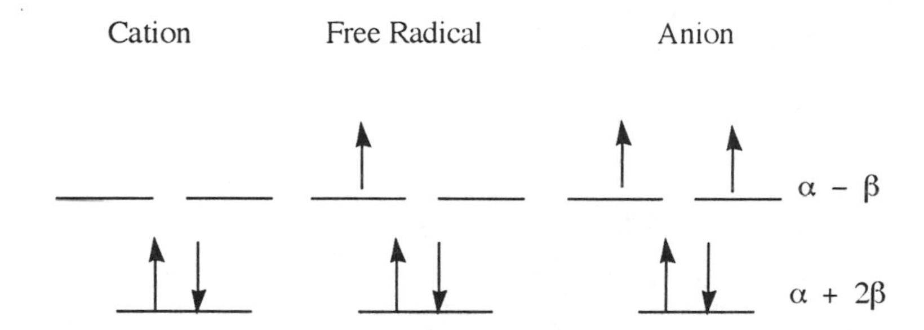

The delocalization energy for the cation is 2β, in contrast to the anion, which has no delocalization energy. The two orbitals of the same energy, $\alpha - \beta$, are said to be degenerate. In adding electrons to the system, Hund's rule applies, and one electron will have to go into each orbital before doubling up in either can occur. Such a structure for the anion would be an unstable triplet diradical state, according to simple Hückel theory.

The determination of the coefficients for degenerate orbitals is more complex than for nondegenerate orbitals, because there is an infinite variety of combinations of the AOs that may be used. Some of the tricks employed in meeting this situation are shown below.

For Ψ_1 the easiest procedure for this simple molecule is the method of elimination used for ethylene:

$x = -2$

$$C_1x + C_2 + C_3 = 0$$
$$C_1 + C_2x + C_3 = 0$$
$$C_1 + C_2 + C_3x = 0$$

After substituting $x = -2$ and solving for the relationship among the three coordinates, application of the normalization condition leads to

$$C_1 = C_2 = C_3 = 1/\sqrt{3}$$

$$\Psi_1 = \frac{1}{\sqrt{3}}(\psi_1 + \psi_2 + \psi_3)$$

But $x = 1$ gives only $C_1 + C_2 + C_3 = 0$. One of two assumptions is usually made at this point: set one coefficient equal to zero or set two coefficients equal to each other. For Ψ_2 let us assume $C_1 = C_2$; then

$$C_1^2 + C_2^2 + C_3^2 = 1$$

$$C_1^2 + C_1^2 + (-2C_1)^2 = 1$$

$$6C_1^2 = 1$$

$$C_1 = 1/\sqrt{6}$$

and

$$\Psi_2 = 1/\sqrt{6}\psi_1 + 1/\sqrt{6}\psi_2 - 2/\sqrt{6}\psi_3$$

To find the coefficients of Ψ_3, we will construct a table of C values and make use of the fact that the sum of all the C_i^2 values must equal one and so on for the C_2 and C_3 values.

	C_1	C_2	C_3
Ψ_1	$1/\sqrt{3}$	$1/\sqrt{3}$	$1/\sqrt{3}$
Ψ_2	$1/\sqrt{6}$	$1/\sqrt{6}$	$-2/\sqrt{6}$
Ψ_3	a	b	c

$$(1/\sqrt{3})^2 + (1/\sqrt{6})^2 + (a)^2 = 1$$

$$a = \pm 1/\sqrt{2}$$

$$(1/\sqrt{3})^2 + (1/\sqrt{6})^2 + (b)^2 = 1$$

$$b = \pm 1/\sqrt{2}$$

$$(1/\sqrt{3})^2 + (-2/\sqrt{6})^2 + (c)^2 = 1$$

$$c = 0$$

These MOs are diagramed below:

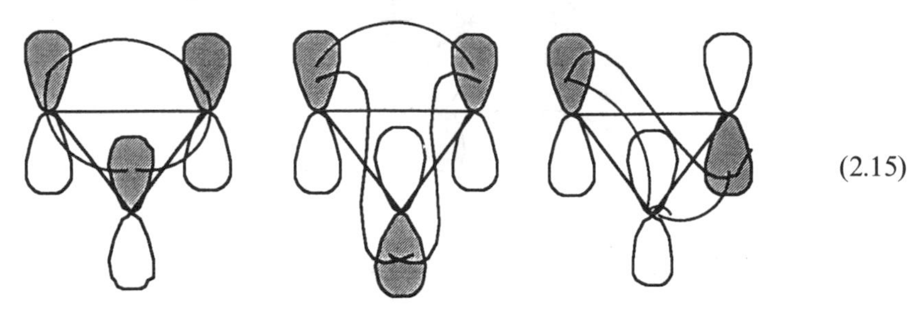

(2.15)

2.8 Benzene

Because of the importance of benzene as the prime example of the aromatic hydrocarbons, we give here the secular determinant and wave functions for the molecule. The six π electrons occupy the lowest three energy levels, giving an E_π of $6\alpha + 8\beta$. When the benzene E_π is compared with a localized cyclohexatriene whose energy is that of three ethylenes, one calculates

$$\begin{vmatrix} x & 1 & 0 & 0 & 0 & 1 \\ 1 & x & 1 & 0 & 0 & 0 \\ 0 & 1 & x & 1 & 0 & 0 \\ 0 & 0 & 1 & x & 1 & 0 \\ 0 & 0 & 0 & 1 & x & 1 \\ 1 & 0 & 0 & 0 & 1 & x \end{vmatrix} = 0$$

$$x = -2, -1, -1, 1, 1, 2$$

a DE of 2β. The molecular orbitals and wave functions are depicted in ascending order of energy in Figure 2.1. Note that the wave functions Ψ_2 and Ψ_3 are degenerate, as are Ψ_4 and Ψ_5.

2.9 Cyclobutadiene

Cyclobutadiene is a molecule with a long and honorable history. In part this is because one can write two equivalent structures, analogous to the two Kekule structures for benzene. Yet, unlike benzene, cyclobutadiene remained an unknown molecule, a target for synthetic and mechanistic organic chemistry until 1965, when its existence was first demonstrated by Petit and co-workers at the University of Texas. The secular determinant for cyclobutadiene resembles that for butadiene given previously, except now the off-diagonal elements representing the interactions of atoms 1 and 4 must be included. Solutions of the quadratic secular equation give roots of -2, 0, 0, and 2. The with spectral properties and chemical reactivity. Multiplicity is given by the term $(2S + 1)$, where S is given as the sum of the electron spin assignments in the following sense. When the electrons are placed in antiparallel pairs, $S = [2(+1/2 - 1/2) + 1] = 1$, and the molecule is a closed shell ground state singlet. With only one unpaired spin, $S = [2 \times 1/2 + 1] = 2$, a doublet state characteristic of a free radical. In the case of cyclobutadiene above, the electrons are placed according to Hund's rule and the multiplicity is $[2(1/2 + 1/2) + 1] = 3$, which corresponds to a triplet ground state. Such a state corresponds to a diradical and is expected to be very chemically reactive. The simple HMO treatment of cyclobutadiene pictures a square molecule of no delocalizing stabilization and diradical character.

This oversimplistic view is corrected by evoking the *Jahn-Teller effect*, which states that molecules with such degenerate orbitals may remove the degeneracy by distorting the geometry of the π system, thereby lifting the degeneracy.

The degeneracy of orbitals 2 and 3 disappears if the molecule is allowed to become rectangular. This possibility is realized in more advanced MO treatments. The chemical properties observed for cyclobutadiene are consistent with a rectangular structure that is a highly strained diene with no triplet character.

2.10 Heteroatom Systems

The binding energy for a carbon $2p$ electron is given by the Coulomb integral α. It goes without saying that the value of α will vary with the electronegativity of the carbon, and this, in turn, is affected by changes in hydridization or by substituents attached to the carbon. In the simplified HMO method previously considered, all values of α in a carbon π system were taken as being the same.

Similarly, the assumption is overly naive that all values of β for carbon–carbon bonds are the same in a given π system. It is known that β varies with bond length, and various means of incorporating this variation have been attempted. Thus, two values of β should be used even for butadiene. However, the computational difficulties of introducing such variations quickly magnify even HMO calculations to proportions requiring a computer.

When heteroatoms are introduced in a π system, new values of α and β are called for. Ideally, one would like to calculate these from first principles, but more realistically, one can drive empirical values more readily. This is done by expressing the values as a function of the values for carbon, i.e.,

$$\alpha_X = \alpha + h\beta$$

$$\alpha_{CX} = k\beta$$

A variety of values of h and k for heteroatoms have appeared in the literature, and these have been reviewed by Streitweiser.[3] Because the values depend on whether the heteroatom contributes one or two electrons to the π system, two sets of values for heteroatoms such as nitrogen and oxygen are usually given:

$$\alpha_N(1) = \alpha + 0.5\beta$$

$$\alpha_N(2) = \alpha + 1.5\beta$$

$$\beta_{C-N} = 0.8\beta$$

$$\beta_{C=N} = \beta$$

$$\alpha_O(1) = \alpha + \beta$$

The secular determinant for methylene imine, $CH_2{=}NH$, will be analogous to that for ethylene:

$$\begin{vmatrix} \alpha - E & \beta \\ \beta_{CN} & \alpha_N(1) - E \end{vmatrix} = \begin{vmatrix} \alpha - E & \beta \\ \beta & \alpha + 0.5 - E \end{vmatrix} = 0$$

Defining $(\alpha - E)/\beta = X$ gives

$$\begin{vmatrix} x & 1 \\ 1 & x + 0.5 \end{vmatrix} = 0$$

and

$$x^2 + 0.5x - 1 = 0$$

$$x = -1.28,\ 0.78$$

$$E = \alpha + 1.28\beta, \text{ and } \alpha - 0.78\beta$$

$$E_\pi = 2\alpha + 2.56\beta$$

which differs from the value of ethylene by 0.56β.

Calculation of the coefficients proceeds as before to give the following MOs. Note the effect of the polarization due to the more electronegative nitrogen.

$$\Psi_1 = 0.62\psi_1 + 0.78\psi_2$$

$$\Psi_2 = 0.78\psi_1 - 0.62\psi_2$$

Remember, for ethylene, $\Psi = 0.707\psi_1 \pm 0.707\psi_2$.

For pyridine it can be shown that the three occupied orbitals have energies

$$E_1 = \alpha + 2.11\beta$$

$$E_2 = \alpha + 1.17\beta$$

$$E_3 = \alpha + 1.00\beta$$

$$E_{\text{total}} = 6\alpha + 8.56\beta$$

The HMO values of the net charge density for aniline and nitrobenzene are shown for comparison. As far as the π systems are concerned, aniline and the benzyl anion are isoelectronic. Nitrobenzene resembles the benzyl cation in having a positive formal charge on the first atom removed from the ring. To a first approximation, the charges of the benzyl species can be taken as examples for charge expectations in the two nitrogen-containing entities. The values for aniline and nitrobenzene were obtained from the program HMO referred to in the preceding chapter.

In the salt formation reactions of aromatic amines, the amine nitrogen is protonated, removing the lone electron pair on nitrogen from interaction with the π system of the ring. As substituent effects cause the lone pair to become more deeply involved in conjugative interactions with the ring, the amine nitrogen will become increasingly less basic in character. The net π charge at this position can be determined within the limits of perturbation molecular orbital (PMO) theory as proportional to those on the appropriate hydrocarbons.

The concepts of bond order and charge densities, of course, apply equally well to heteroatom systems. Returning to methylene imine, one can write the resonance-contributing structures as

$$CH_2{=}NH \leftrightarrow CH_2^+{-}\ddot{\bar{N}}H$$

a result which is consistent with the MO coefficients given previously. The π bond order p is 0.96; the charge density is $+0.22$ on carbon and -0.22 on nitrogen.

For pyridene the charge densities are as shown below. Here q_N is given by (1-ED), since the nitrogen, like the carbons, contributes only one electron to the π system.

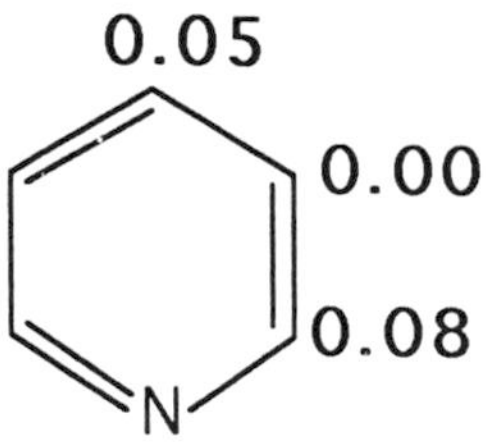

0.05
0.00
0.08
N

2.13 Dipole Moments

For non-AHCs the ED will generally vary from unity. Therefore, non-AHCs will often have dipole moments due to π charge disparities around the molecule. To assume no contribution to the molecular dipole moment as due to the underlying σ framework is an obvious oversimplification. Clearly, polarization of the π system will be reflected in the σ system as well. But often the σ contribution appears to be small, and dipole moments calculated from π charges are at least "in the ballpark." Remembering that dipole moments are vector quantities, we may choose some arbitrary origin and calculate the contribution of each carbon to the total moment.

The charge densities for methylenecyclopropene are as follows:

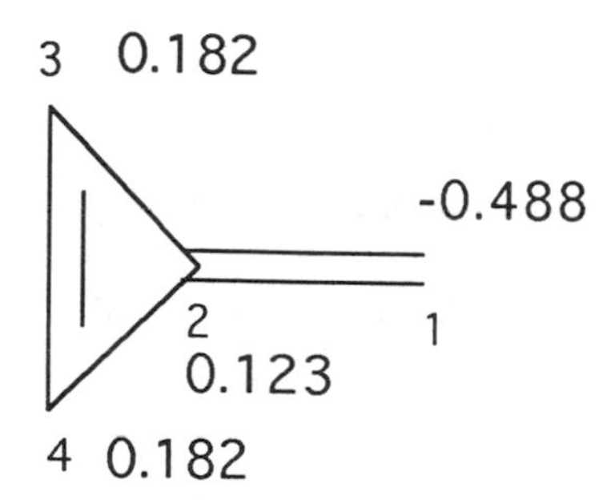

This example is chosen as the simplest non-AHC. Unfortunately, its dipole moment has not been experiementally measured. For simplicity we assume all bond lengths to be 1.40 Å. The dipole moment associated with a given bond is a vector whose scalar value is equal to the product of the bond length and the charge at the end of the bond. If atom 2 is chosen as the origin, the contribution involving the charge on atom 2 goes to zero. For the shake of simplicity, it will be assumed that all bonds are 1.40 Å in length.

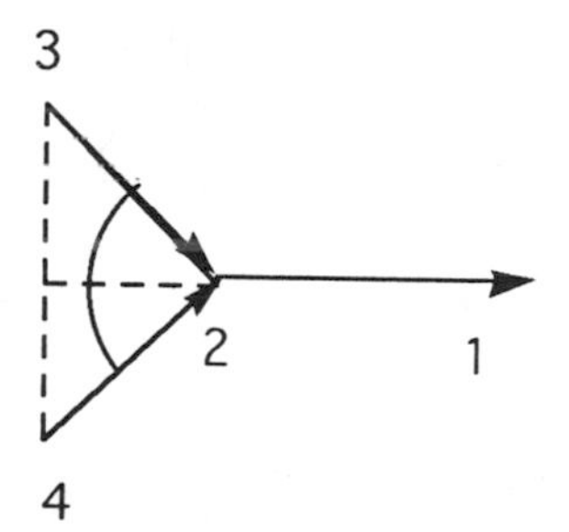

The internal angle shown is 60°

The definition of a bond dipole moment (μ) is the product of a bond distance vector (d) and the charge (e), and the component due to C-2 is nil (i.e., $d = 0$). The vector sum is then given by

$$\mu = 4.77[(0.488)(1.40) + 2(0.182)(1.40)\cos 30°] = 5.37\text{D}$$

where the factor 4.77 is the appropriate conversion to Debye units.

Dipole moments calculated for non-AHCs by the simple HMO method are invariably too large. Experience with less severe approximations (and therefore more sophisticated calculations) shows that the HMO method always exaggerates the charge densities in non-AHCs, with the resultant enhancement of the dipole moments in these compounds. Some idea of the magnitude of the error is given in problem 7, where results from two types of calculations are compared for fulvene.

Problems

1. Write the secular determinant for methylenecyclopropene.

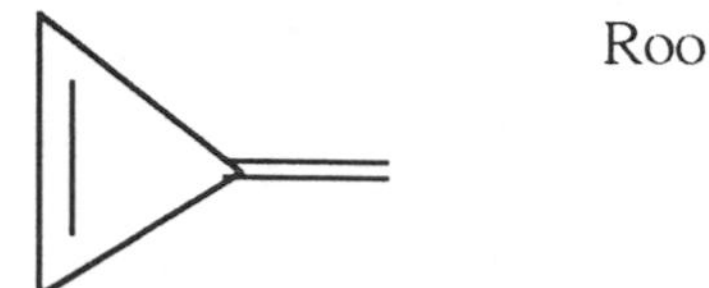

Roots = -2.170, -0.311, 1.00 and 1.481

Given the roots above, calculate E_π and DE.

2. The π energy for benzene is to be compared to that of a nondelocalized cyclohexatriene. Write the secular determinant for this molecule and calculate E_π and DE for benzene.

3. How would the secular determinants of the following differ?

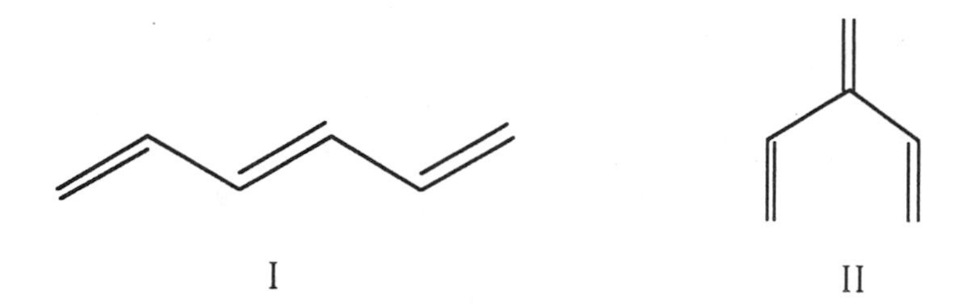

Given the roots below for each structure, calculate DE for each. How does cross conjugation in II affect the stability of the hexatriene system?

I $x = \pm 1.802, \pm 1.247, \pm 0.445$

II $x = \pm 1.932, \pm 1.000, \pm 0.518$

4. Write the secular determinant for cyclobutadiene, find the roots, and calculate the DE and wave functions for the molecule. Diagram the wave functions.

5. Continuing with problem 4, calculate the bond orders and charge densities for cyclobutadiene using the wave functions developed in 4.

6. Given below the roots and wave functions for the three lowest filled orbitals in fulvene, calculate DE and the values of q and p.

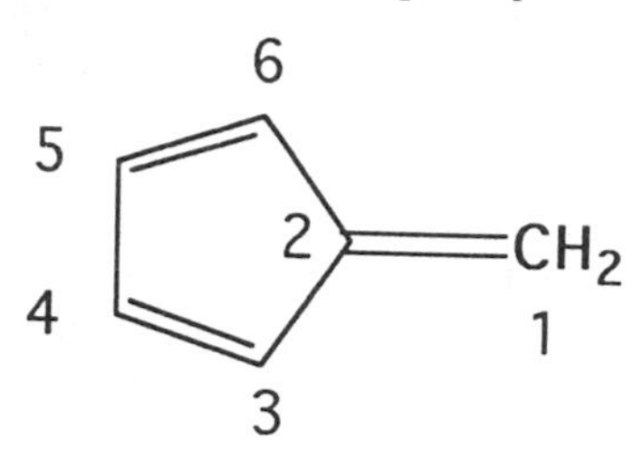

	ψ_1	ψ_2	ψ_3	ψ_4	ψ_5	ψ_6
Ψ_1	0.245	0.523	0.430	0.385	0.385	0.430
Ψ_2	0.500	0.500	—	−0.500	−0.500	—
Ψ_3	—	—	0.601	0.372	−0.372	−0.601

$X = -2.115, -1.000, -0.618$

7. The diagram below gives the dimensions of the fulvene molecule (derived from the x-ray structure of dimethylfulvene). Calculate the dipole moment using the HMO charge densities from exercise 2 and from the self-consistent field values given below. A value estimated from dimethylfulvene is 1.2D; a microwave determination gives an experimental value of 0.44D.

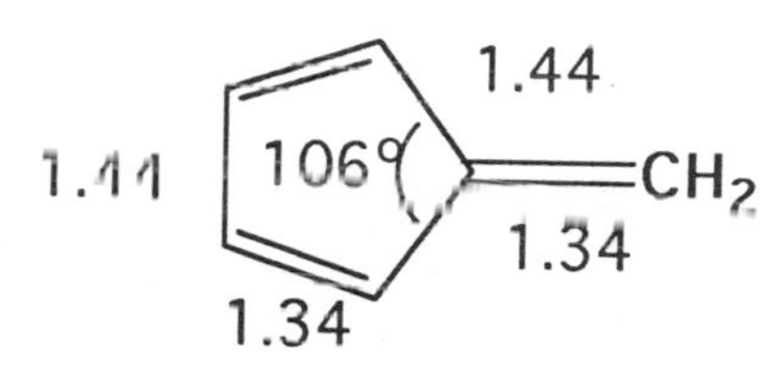

8. Calculate the electron energy, wave functions, bond order, and charge densities for formaldehyde. Compare these entities with those for ethylene and methylene imine.

9. Show that −2.686, 0.186, and 1.000 are satisfactory roots for azacyclopropene. Determine the wave functions, bond orders, and charge densities for this system. Compare these with the cyclopropenyl anion (wave functions given on p. 22).

10. Given the following set of wave functions for phenol (filled orbitals only), calculate the EDs and charge densities for all atoms.

Atom Number

Wave Functions	1	2	3	4	7
Ψ_1	0.603	0.312	0.181	0.144	0.597
Ψ_2	−0.173	0.175	0.456	0.563	−0.420
Ψ_3	0.000	0.500	0.500	0.000	0.000
Ψ_4	0.237	0.450	−0.083	−0.478	−0.545

References

1. M. J. S. Dewar, "The Molecular Orbital Theory of Organic Chemistry," McGraw-Hill, New York, 1969.

2. Trinity Software, P. O. Box 960, Campton, NH 03223 offers a convenient Macintosh HMO program. Several software packages for calculations of this type are available from the Journal of Chemical Education via JCE:Software, Department of Chemistry, University of Wisconsin-Madison, 1101 University Avenue, Madison, WI 53706-1396; E-Mail: JCESOFT@MACC.WISC.EDU. A catalog is available on the Internet at "gopher: JCHEMED. CHEM.WISC:EDU."

3. A. Streitwieser, Jr., *Molecular Orbital Theory for Organic Chemists*, John Wiley & Sons, New York, 1961. A parameterization that upgrades the Hückel treatment is given on page 135. A more modern parameterization is given by F. A. Van-Catledge, *J. Org. Chem.*, **45**, 4801 (1980).

CHAPTER

3

The PMO Method

In HMO calculations, the wave functions for the system are generated by the method of variations applied to a linear combination of the atomic orbitals that make up the system. The resultant wave functions describe the system subject to those limitations inherent in the HMO assumptions. Although the availability of computers has allowed the application of HMO theory (and even more sophisticated treatments) to large π electron systems, attempts have been made over the years to produce a shorthand type of calculation that would still provide valid insights into problems of molecular structure and chemical reactivity. The PMO method was developed by Coulson, Longuet-Higgins, and Dewar.[1] It offers not only a rapid means of doing certain types of calculations, but also some surprising knowledge about the meanings of resonance theory, the relative stabilities of hydrocarbons, the effects of hetero-atoms and substituents, and the factors influencing certain types of chemical reactions.

As an introduction to the subject, consider a complex new system on which one would like to do calculations of the type covered in the preceding chapters. Suppose the wave functions for similar systems are already known. The solutions for the new system might be considered to arise from a perturbation of the latter. A case in point is the relation of the cyclopropenyl anion and azacyclopropene in the last chapter. Had we set the value of h in the Coulomb integral for nitrogen equal to zero, the secular determinants for both molecules would have become the same. The wave functions for azacyclopropene reflect the perturbing effects of the inductive effect of the nitrogen, and this is reflected also in the energy levels and charges in the molecule.

3.1 First-Order Perturbation Theory

The following treatment of first-order perturbation theory provides much of the approach used to develop the PMO method and, thus, provides the background for what is to come. For the perturbed system, let us write the Hamiltonian as

$$H = H^0 + \lambda H^1$$

where H^0 is the Hamiltonian for the unperturbed system and λH^1 is the perturbation. When the coefficient λ is zero, the perturbation is nil. Let us define $\Psi_1^0, \Psi_2^0, \Psi_3^0 --- E_1^0, E_2^0, E_3^0 ---$ as eigenfunctions and eigenvalues of H^0. Furthermore, we will assume all eigenvalues to be different; the case with degeneracies requires more specialized treatment. Similarly, $\Psi_1, \Psi_2, \Psi_3 ---$ $E_1, E_2, E_3 ---$ characterize the values for the perturbed system. The corrections to the zeroth-order terms are $\Psi_1^1, \Psi_2^1, \Psi_3^1 ---$ and $E_1^1, E_2^2, E_3^3 ---$. Then,

$$H^0\Psi_1^0 = E_1^0\Psi_1^0 \tag{3.1}$$

$$\Psi_i = \Psi_i^0 + \lambda\Psi_1^1 \tag{3.2}$$

$$E_1 = E_i^0 + \lambda E_i^1 \tag{3.3}$$

and

$$H\Psi_i = E_i\Psi_i \tag{3.4}$$

Substituting in Eq. 3.4:

$$(H^0 + \lambda H^1)(\Psi_i^0 + \lambda\Psi_i^1) = (E_{ii}^0 + \lambda E_i^1)(\Psi_i^0 + \lambda\Psi_i^1) \tag{3.5}$$

$$H^0\Psi_1^0 + \lambda H^0\Psi_i^1 + \lambda H^1\Psi_i^0 + \lambda^2 H^1\Psi_i^1$$
$$= E_i^0\Psi_1^0 + \lambda E_i^0\Psi_i^1 + \lambda E_i^1\Psi_i^0 + \lambda^2 E_i^1\Psi_i^1 \tag{3.6}$$

Terms in λ^2 may be dropped from Eq. 3.6, since λ is a small fraction for most perturbations. Then subtracting Eq. 3.1 from Eq. 3.6, we get the first-order perturbation equation 3.7:

$$H^0\Psi_i^1 + H^1\Psi_1^0 = E_i^0\Psi_i^1 + E_i^1\Psi_i^0 \tag{3.7}$$

However, Eq. 3.7 still does not solve the problem, as the corrective function Ψ^1 is not known. It is usual to approximate this corrective wave function as a linear combination of the unperturbed wave functions, thus

$$\Psi_i^1 = \sum a_{ij}\Psi_j^0$$

Substituting in Eq. 3.7:

$$H^1\Psi_i^0 + \sum_j a_{ij}H^0\Psi_j^0 = E_i^1\Psi_i^0 + E_i^0 \sum_j a_{ij}\Psi_j^0 \tag{3.8}$$

Multiplying both sides by Ψ_k^0 and integrating:

$$\int \Psi_k^0 H^1\Psi_i^0\, d\tau + \sum a_{ij} \int \Psi_k^0 H^0\Psi_j^0\, d\tau = E_i^1 \int \Psi_k^0\Psi_i^0\, d\tau + E_i^0 \sum a_{ij} \int \Psi_k^0\Psi_j^0\, d\tau \tag{3.9}$$

and

$$\int \Psi_k^0 H^1\Psi_i^0\, d\tau + \sum_j a_{ij}E_j^0\,\delta_{jk} = E_i^1 \int \Psi_k^0\Psi_i^0\, d\tau + E_i^0 \sum a_{ij}\delta_{kj} \tag{3.10}$$

When $j = k$, Eq. 3.10 reduces to

$$\int \Psi_k^0 H^1\Psi_i^0\, d\tau + a_{ik}E_k^0 = E_i^1\delta_{ik} + E_i^0 a_{ik} \tag{3.11}$$

Now if $i = k$, Eq. 3.11 further reduces to

$$\int \Psi_i^0 H^1\Psi_i^0\, d\tau = E_i^1 \tag{3.12}$$

Thus, the perturbation correction to E_1 does not depend on a knowledge of the perturbed wave function, but depends only on a knowledge of the perturbation operator.

When $i \neq k$, Eq. 3.11 reduces to

$$\int \Psi_k^0 H^1\Psi_i^0\, d\tau + a_{ik}(E_k^0 - E_i^0) = 0 \tag{3.13}$$

and

$$a_{ik} = \frac{\int \Psi_k^0 H^1\Psi_i^0\, d\tau}{(E_k^0 - E_i^0)} \tag{3.14}$$

The coefficients a_{ii} cannot be determined by Eq. 3.14, since E_i^0 will equal E_k^0. When all other coefficients have been determined, however, normalization may be employed to determine a_{ii}.

The perturbed wave equation will be of the form

$$\Psi_i = \Psi_i^0 + \sum_j \frac{\int \Psi_i^0 H^1 \Psi_i^0 \, d\tau}{(E_i - E_j^0)} \Psi_j^0 \tag{3.15}$$

3.2 Internal Changes of α and β

Let us consider what happens to the energies of a conjugated π system when a heteroatom replaces a carbon. Rather than making the usual approximations of Chapter 4 and solving new sets of determinants, a perturbational approach will be used. Furthermore, the definitions used above will be continued, except for a convenient change in the subscripted variable to be consistent with our earlier expressions for the LCAO, i.e., for the u-th MO,

$$\Psi_u = \sum_i c_{ui} \psi_i$$

Reiterating from above,

$$E_u = E_u^0 + \lambda E_u^1$$

and from Eq. 3.12,

$$E_u^1 = \int \Psi_u^0 H^1 \Psi_u^0 \, d\tau$$

For convenience, λ will be taken inside H^1, leading to

$$\delta E_u = E_u^1 - E_u^0 = \int \Psi_u^0 H^1 \Psi_u^0 \, d\tau \tag{3.16}$$

From our earlier definitions of the Coulomb and resonance integrals of carbon atom i,

$$\alpha_i = \int \psi_i H^0 \psi_i \, d\tau$$

$$\beta_{ij} = \int \psi_i H^0 \psi_j \, d\tau$$

and for the perturbed functions

$$A_i = \int \psi_i H \psi_i \, d\tau$$

$$B_{ij} = \int \psi_i H \psi_j \, d\tau$$

The correction terms then become

$$\delta\alpha_i = A_i - \alpha_i = \int \psi_i H^1 \psi_i \, d\tau$$

$$\delta\beta_{ij} = B_i - \beta_i = \int \psi_i H^1 \psi_j \, d\tau$$

Now Eq. 3.16 can be rewritten as

$$\delta E_u = \int (\sum c_{ui}\psi_i) H^1 (\sum C_{ui}\psi_i) \, d\tau$$

$$= \sum c_{ui}^2 \int \psi_i H^1 \psi_i \, d\tau + \sum_{i \neq j} \sum c_{ui} c_{uj} \int \psi_i H^1 \psi_j \, d\tau \qquad (3.17)$$

Inserting our definitions above,

$$\delta E_u = \sum_i c_{ui}^2 \delta\alpha_i + 2 \sum_{i<j} \sum c_{ui} c_{uj} \delta\beta_{ij} \qquad (3.18)$$

The factor of 2 appears in the last equation because $\delta\beta_{ij} = \delta\beta_{ji}$, as shown in Chapter 1:

$$\delta E = \sum_i \mathrm{ED}_i \delta\alpha_i + 2 \sum_{i<j} \sum p_{ij} \delta\beta_{ij}$$

For the total energy change due to alterations in α and β (each orbital containing two electrons),

$$\delta E = 2 \sum_u^{\mathrm{occ}} E_u^1 - 2 \sum_u^{\mathrm{occ}} E_u^0 = 2 \sum_u^{\mathrm{occ}} E_u = 2 \sum^{\mathrm{occ}} (\sum c_{ui}^2 \delta\alpha_i + 2 \sum_{i<j} \sum c_{ui} c_{uj} \delta\beta_{ij}) \qquad (3.19)$$

From our previous definitions of the electron density ED and bond order P_{ij}, Eq. 3.19 may be rewritten as

$$\delta E = \sum_i \mathrm{ED}_i \delta\alpha_i + 2 \sum_{i<j} \sum p_{ij} \delta_{ij} \qquad (3.20)$$

We will have several occasions to use this equation again, but we may immediately check one of the consequences of heteroatom substitution seen in the last chapter. If the approximation $\beta_{\mathrm{CN}} = \beta$ is again employed, Eq. 3.20 reduces to

$$\delta E = \sum_i \mathrm{ED}_i \delta\alpha_i \qquad (3.21)$$

Since the ED in our AHC is unity and $\delta\alpha_i = h\beta$, δE is simply $h\beta$. The DE for a heterocycle such as pyridine will be the same as that of the corresponding hydrocarbon, since the methylene imine fragment used in our model compound will also be reduced by $h\alpha$. Of course, this is in agreement with our HMO-derived conclusions for the one specific case.

3.3 Amine Base Strengths

Previously we saw that a qualitative assessment of amine base strengths could be made on the assumption that a CH_2^- group could represent an NH_2 group. The equilibrium between two bases and their conjugate acids may be written as

$$B_1H^+ + B_2 \rightleftharpoons B_1 + B_2H^+$$

If we are concerned with base strengths in aqueous solution, then B_1H^+ would be a water molecule and B_1 the hydroxide ion. The standard free energy change for the reaction is given by

$$\Delta G^0 = -RT \ln K = -2.303\,RT(\mathrm{p}Ka_1 - \mathrm{p}Ka_2)$$

In discussions of relative base strengths of a related series of amines, it is often assumed that electrostatic interactions in solution may be neglected and that ionization entropies are about the same. If so, then the free energy change contains only potential energy terms related to the change in π electron energies. Given B_1 as some reference base, the above equation can be rewritten as

$$2.303\,RT\,\mathrm{p}Ka = \text{constant} - (E_{\mathrm{BH}^+} - E_{\mathrm{B}}) \tag{3.22}$$

Now for the monoaza substituted heterocyclics such as pyridine, quinoline, and isoquinoline, the total π energy from Eq. 3.21 above is

$$E_{\mathrm{B}} = E^0 + h_{\mathrm{N}}\beta$$

In the conjugate acid, the parameter h_{N^+} will assume some value larger than h_{N} to express the enhanced electronegativity of the positively charged nitrogen:

$$EB_{\mathrm{H}^+} = E^0 + h_{\mathrm{N}^+}\beta$$

The change in π energy on protonation then reduces to

$$(EB_{\mathrm{H}^+} - E_{\mathrm{B}}) = (h_{\mathrm{N}^+} - h_{\mathrm{N}})\beta$$

Thus, all the monoaza heterocyclics are predicted by PMO theory to have the same base strength. In point of fact, data for a number of these compounds show p*K*a values in the range of 4.2 to 5.2. Acridine is the exception, having a value of 5.6 in water.

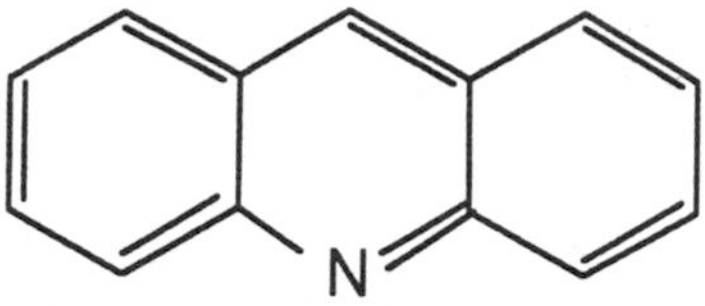

Acridine

The p*K*a is comparable with the others in a 50% aqueous alcohol solvent.

Spectroscopic and NMR studies have shown that amine-substituted heterocyclics such as the two below protonate on the ring nitrogen first.

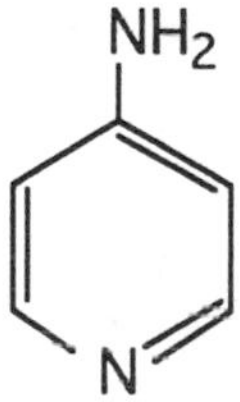

4-Aminopyridine

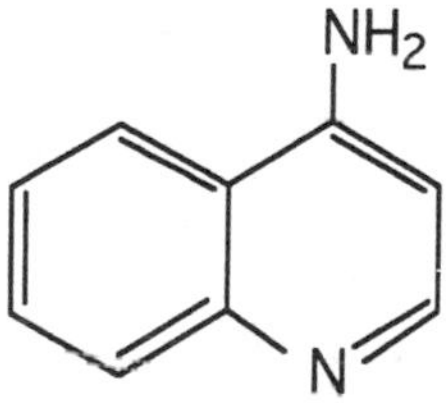

4-Aminoquinoline

The p*K*a values of a series of such compounds have been examined, and the range of values is considerably larger than for the heterocyclics above, the p*K*a being influenced by the presence and orientation of the amine group on the ring. The energy changes in the 4-aminoquinoline system may be summarized as follows [note the use of Eq. 3.21 implied in (b) and (d)]:

$h_N \beta$

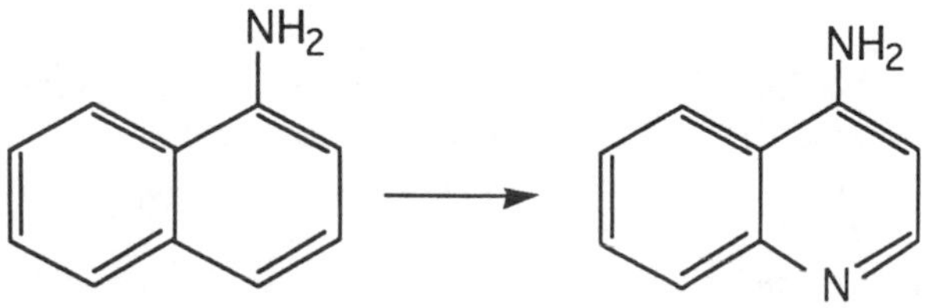

$ED_4 \, h_N \beta$

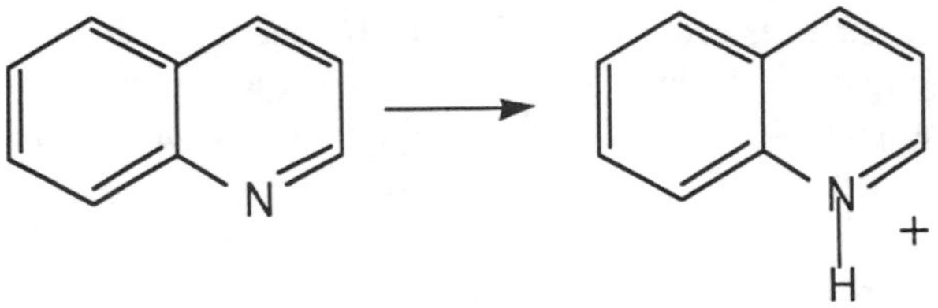

$(h_{N+} - h_N)\beta$

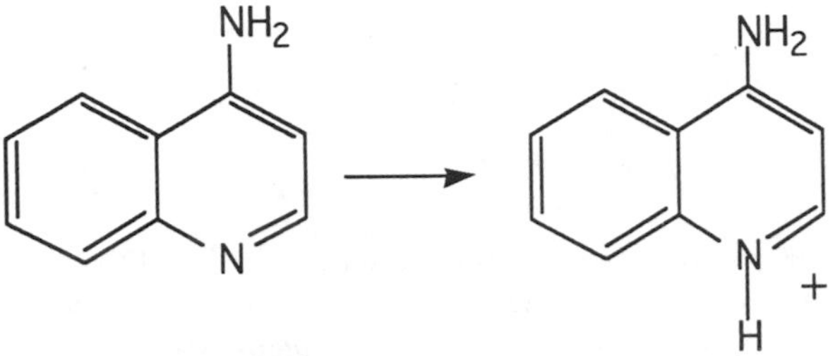

$ED_4(h_{N+} - h_N)\beta$

The electron density ED_4 can be obtained from the square of NBMO coefficients as shown in Chapter 2. From Eq. 3.22 above, then

$$2.303\,RT\,\mathrm{p}Ka = \text{constant} - ED_4(h_{N^+} - h_N)\beta \tag{3.23}$$

Since ED_4 and $(h_{N^+} - h_N)\beta$ are positive numbers while β is negative, an increase in ED_4 will lead to an increase in p*K*a.

Longuet-Higgins in 1950 applied the above scheme to some 20 amine heterocyclics. A least squares fit of his listed data gives a correlation coefficient of 0.911 (a perfect linear fit is 1; no fit is zero). The two poorest points on the curve are due to 1-aminoacridine and 9-aminophenanthridine:

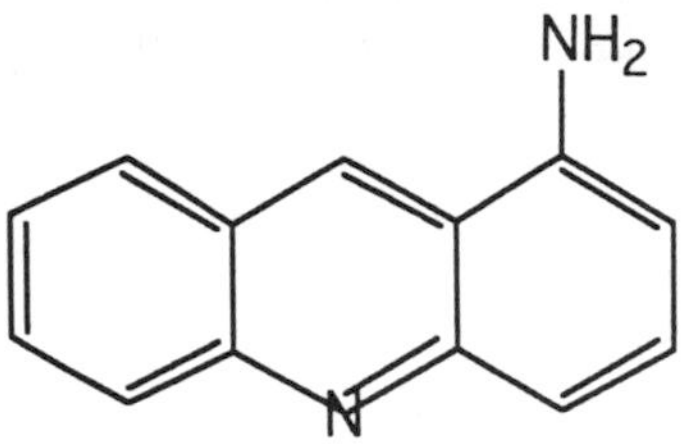

1-Aminoacridine

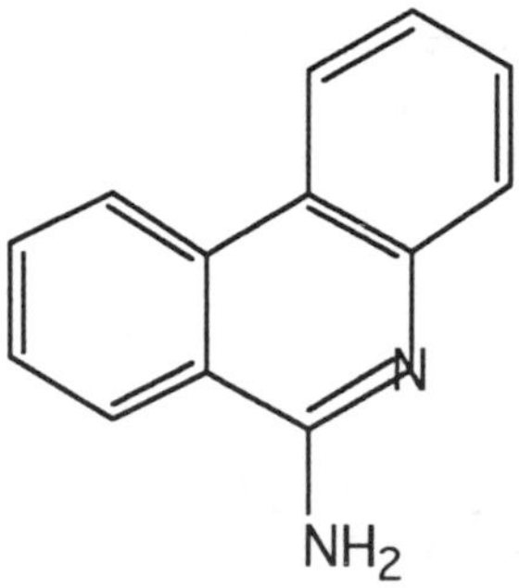

9-Aminophenanthridine

It has been proposed that *peri* interactions and an *ortho* steric hindrance, respectively, operate to produce these aberrations. When these two points are omitted from the plot, the correlation coefficient is improved to a quite satisfactory value of 0.959. For this plot the slope $(h_{N^+} - h_N)\beta/2.303\,RT$ is 17.57. Given the usual thermochemical value for β (18–20 kcal/mol), one calculates that $(h_{N^+} - h_N)$ is in the range of 1.2 to 1.30.

3.4 The Union of π-Systems

The secular determinant for classical butadiene (Chapter 2) differs from that for the delocalized system in the absence of the off-diagonal elements for the resonance integrals between carbons 2 and 3. As far as classical butadiene is concerned, the π system is equivalent to two isolated ethylenes. The union of two ethylenes to form delocalized butadiene represents a perturbation in which β_{23} goes from zero to some finite value.

Just as in our earlier examination of first-order perturbation theory, systems with degenerate energy levels cause some difficulties, and our initial considerations will be directed to systems without such degeneracies, i.e., we will avoid the case of the union of two ethylenes for the moment. Consider two isolated π systems, R and S, which undergo union to form a delocalized system R—S. Let us define the following:

$$H^{RS} = H^{R} + H^{S} + H^{1}$$

where H^{RS} is for the delocalized whole composed of H^{R} for R, H^{S} for S, and H^{1} for the perturbation produced by the union.

For R:

$$\Psi_m = \sum_i a_{mi}\psi_i$$

$$E_m = \int \Psi_m H^{R} \Psi_m \, d\tau$$

For *S*:

$$\Phi_n = \sum_j b_{nj}\phi_j$$

$$F_n = \int \phi_j H^{S} \phi_j \, d\tau$$

By the orthogonality-normality conditions:

$$\int \Psi_m \Psi_n \, d\tau = \delta_{nm}$$

$$\int \Phi_m \Phi_n \, d\tau = \delta_{mn}$$

$$\int \Psi_m \Phi_n \, d\tau = 0$$

Also

$$\int \Psi_m H^1 \Psi_n \, d\tau = \int \Phi_m H^1 \Phi_n \, d\tau$$

and

$$\int \Psi_m H^1 \Phi_n \, d\tau = a_{mr} b_{ns} \beta_{rs}$$

The latter integral comes about because the union occurs only between atoms r and s, and the new bond is characterized with β_{rs}.

Subject to the proviso of no degeneracies, the perturbations to E_m and F_n are given by first-order perturbation theory as

$$\delta E_m = \int \Psi_m H^1 \Psi_m \, d\tau = 0$$

and

$$\delta F_n = \int \Phi_n H^1 \Phi_n \, d\tau = 0$$

Although we have not considered second-order perturbation theory, extension of our previous considerations to the λ^2 terms leads to a second-order correction to the energy of the form

$$\delta E_m^{(2)} = \sum_n \frac{a_{mr}^2 b_{ns}^2 \beta_{rs}^2}{E_n - F_n}$$

$$\delta F_n^{(2)} = \sum_m \frac{a_{mr}^2 b_{ns}^2 \beta_{rs}^2}{F_n - E_m}$$

In view of the fact that the bonding MOs of AHCs do not vary greatly in energy, it has been suggested that this second-order correction might be simplified by using some mean value of $\bar{E}$ for the energy difference terms leading to

$$\delta E^{(2)} = E_{\mathrm{R+S}} + \frac{\beta_{\mathrm{RS}}^2}{2\bar{E}}$$

Although the first-order result above may appear disappointing, in the subsequent chapter we will see that the result appears to correspond to reality.

We can demonstrate the first-order approximation by considering the union of ethylene with butadiene to form benzene. Since there will be no perturbational correction, the energy of the π system will simply be the sum of the energies of the parent systems, i.e., $2(\alpha + \beta)$ from the ethylene and $2(\alpha + 1.618\beta) + 2(\alpha + 0.618\beta)$ from the butadiene for a total of $(6\alpha + 6.472\beta)$. This figure is to be compared with the HMO value of $(6\alpha + 8\beta)$.

When the stricture regarding degeneracy is dropped, the first-order change in energy ceases to be zero. The case of interest regards the union of two odd-AHC fragments. The change in π energy in this case is

$$\delta E = 2a_{or}b_{os}\beta = 2p_{rs}\beta \tag{3.27}$$

where a_{or} and b_{os} are the NBMO coefficients of the r atom in R and the s atom in S. Equation 3.27 may be illustrated by the union of methyl with benzyl to form styrene:

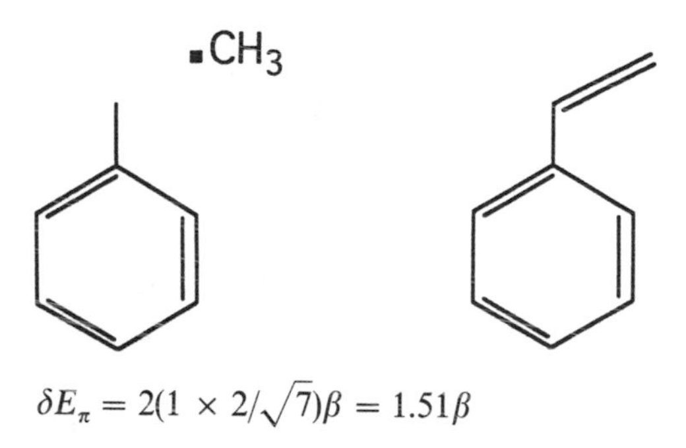

$$\delta E_\pi = 2(1 \times 2/\sqrt{7})\beta = 1.51\beta$$

In this case, the single carbon fragment has a coefficient 1 due to normalization, whereas the coefficient for the external carbon NBMO was derived in Chapter 2. The HMO value for E for benzyl is $7\alpha + 8.72\beta$, and that for styrene is $8\alpha + 10.42\beta$. Thus, the HMO value for δE_π is 1.70β (the energy of the electron localized on the one-carbon fragment being just α). Whether one should consider this good agreement is a matter of opinion. It is certainly evident that the calculations in Chapter 2 and Eq. 3.27 are much easier than finding the roots for a 7×7 and an 8×8 pair of determinants.

The above treatment of PMO theory is, at best, an introduction. Dewar (1952) and others have greatly extended the method, and in the chapters to come we shall state many of these results without going into the detailed derivations.

Problems

1. According to PMO theory, the total π energy for the union of two hydrocarbons is

$$E_\pi = E_{\mathrm{R}} + E_{\mathrm{S}} + E^1$$

where E^1 is the perturbation energy. For AHCs the value of E^1 was shown to be zero. Calculate E for butadiene assuming that (1) it is formed by the union of two ethylenes and (2) it is formed by the union of methyl with allyl. Compare these results with the HMO value.

2. Hexatriene can be considered to be formed by the union of two allyls or by the union of methyl with pentadienyl. Calculate E_π by each path.

CHAPTER

4

Aromaticity, Antiaromaticity, and Resonance

4.1 General Comments

It has been well over a hundred years since Kekule, struggling to reconcile the molecular formula for benzene with its peculiar chemical stability, suggested the ring structure with alternating single and double bonds. It was soon recognized that benzene, naphthalene, anthracene, and related structures all possessed the common traits of stability to oxidation and reluctance to undergo hydrogenation and halogen addition reactions, but a willingness to undergo electrophilic substitution reactions.

In the intervening years, chemists have played the game of attempting to use one set of characteristics or another to define this concept of aromaticity or aromatic character. A variety of physical properties have been suggested at one time or another. Among these may be listed the absorption of light at wavelengths longer than those found for olefins, the anisotropy of magnetic susceptibility of molecular crystals, and the downfield position of the NMR chemical shifts of protons attached to the carbon ring system. None of these have proved totally satisfactory.

As the concept of the covalent electron bond matured during the early part of this century, Robinson introduced the concept of the "aromatic sextet," i.e., six unsaturation electrons in a ring, as a structural feature of aromatic systems. Pauling and Wheland in the 1930s verbalized the constructs of valence bond theory to produce the pictorial resonance theory, which still appears in most sophomore organic chemistry textbooks. According to this approach, conjugated molecules are to be represented as the superpositon of structures written according to the rules of valence.

The above device represented a step forward for the average chemist at work at the laboratory bench. He could "understand" why the bond lengths in butadiene were different from those in unconjugated dienes, why 1,4 additions occurred, and why the heat of hydrogenation was less than expected on the basis of ethylene.

The development of HMO theory provided a more quantitative picture for unsaturated systems. The problem is that simple HMO calculations uniformly exaggerate the extent of electron delocalization and the magnitude of delocalization energies in conjugated systems, whether linear or cyclic. Because of this, there is a failure in the usual molecular additivity properties, such as bond lengths, bond energies, bond moments, heats of atomization, etc.

Experimentally, resonance or delocalization energies are calculated by comparison of heats of combustion or heats of hydrogenation between the conjugated system and some standard "localized" system, usually ethylene or cyclohexene. During the course of certain self-consistent field (SCF) calculations of a type considerably more sophisticated than the HMO variety, Dewar and his co-workers realized that, in fact, the collective properties of conjugated polyenes could be expressed as additive functions of newly defined quantities characterizing the carbon double and single bonds in these structures. In contrast to benzene, for which two classical Kekule structures may be written, butadiene has only one classical structure:

$$CH_2{=}CH{-}CH{=}CH_2$$

The bonds here are referred to as essential double and single bonds. According to HMO calculations, butadiene and longer-chain polyenes have relatively high DEs, and furthermore, the bonds near the center of such molecules tend to assume uniform intermediate lengths and bond orders. The latter observation is not found in the SCF calculations; the double and single bond patterns continue to alternate down the chain. Thus, if one has a characteristic value for the essential double and single bonds, the thermodynamic properties of the polyenes may be calculated quite well.

Dewar contends that the flaw lies in our concept of the localized bond. Initially conceived to explain the directedness of the covalent bond and the geometry of molecules, the localized bond is usually considered to exist between only two atoms and to be composed of the overlap of AOs centered on those atoms. In fact, MOs always extend over the whole molecular framework. In the case of butadiene, there is an interaction between the π electrons associated with the two double bonds, but the same type of interaction occurs in all conjugated polyenes. One can, therefore, find collective properties for all of the conjugated polyenes, which should still be considered

Table 4.1. Bond Increments of Hess and Schad

Bond Designation	Bond Type	π Bond Energy (β)
23	$CH2{=}CH{-}$	2.0000
22	$-CH{=}CH-$	2.0699
22	$CH_2{=}C<$	2.0000
21	$-CH{=}C<$	2.1083
20	$>C{=}C<$	2.1716
12	$>CH{-}CH<$	0.4660
11	$>CH{-}C\leqslant$	0.4362
10	$\geqslant C{-}C\leqslant$	0.4358

to consist of an alternating pattern of essential double and single bonds. Dewar maintained that any hydrocarbon whose collective properties follow an additivity principle can be said to contain localized bonds. Butadiene and other polyenes for which only one classical structure can be written fall in this category, though benzene, naphthalene, etc., do not.

Hess and Schaad (Vanderbilt University) have shown that this definition of the localized bond may be derived even from HMO calculations with an accuracy comparable to those achieved by the SCF results. They have derived the table of π bond energies above. On this basis, the energy for butadiene would be

$$E_\pi = 2E_{23} + E_{12} = 4.4660\beta$$

The HMO value is 4.48β so that the DE is only 0.01β at best. For benzene the calculation is

$$E_\pi = 2E_{22} + 3E_{12} = 7.608\beta$$

This is some 0.39β below the HMO value. This new DE for benzene is smaller than older values because the new reference polyene, cyclohexatriene, has a larger binding energy.

4.2 The PMO Method, Localized Bonds, and Aromaticity

In the preceding chapter a brief introduction to the PMO method was given, although many of the details were omitted in the interest of getting on to the results. It was shown, however, that the union of two even-AHCs to form RS resulted in no first-order change in the π energy. As mentioned, the second-order change is $\beta^2/2\bar{E}$, where $\bar{E}$ was a mean value of the AHCs' bonding MOs.

The new bond between atoms r and s must be an essential single bond, for if it were a double bond—remembering that two even-AHCs are undergoing union—the remaining atoms in R and S would be odd in number. This line of argument goes along with our preceding section to the effect that bonds in classical polyenes are localized.

As we have seen, the first-order term is not zero for the union of two odd-AHC radicals. The same is true for the union between two atoms r and s in the same conjugated system. Here too, the energy change is given by

$$\delta E_\pi = 2p_{rs}\beta$$

Now one of the accepted theorems of the PMO method is that the bond orders between atoms of like parity in an even-AHC are zero. Parity in this context means that the atoms under consideration are either all starred or all unstarred.

An interesting application of these principles is the formation of fulvene, a non-AHC system, by either of the two processes shown here.

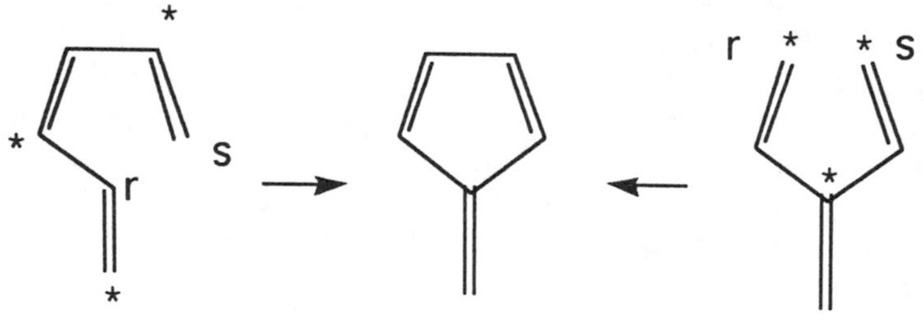

In either case the first-order change in π energy will be zero. Following the same line of thought given for butadiene, either of the newly formed bonds will be essential single bonds. Fulvene is shown by the PMO method to be a cyclic polyene with localized single and double bonds. The HMO DE for fulvene was calculated in Chapter 3, problem 2. The DE derived from the localized bond approach of Hess and Schaad is given as an exercise to follow. Suffice it to say here that the chemistry of fulvene is that of a highly unstable polyene in agreement with the PMO and localized bond results.

The extension of the PMO method to the determination of aromaticity requires the expansion of Eq 3.27 to the case of multiple union, i.e.,

$$E_\pi = 2\sum_{rs} a_{or}b_{os}\beta = 2\sum p_{rs}\beta \tag{4.1}$$

A given cyclic system will be said to be aromatic if its π energy is less than that of the corresponding linear polyene. Thus, we may consider the union of two allyls to give benzene and hexatriene:

$1/\sqrt{2}$ $1/\sqrt{2}$

$-1/\sqrt{2}$ $-1/\sqrt{2}$

$\delta E_\pi = 2\beta$

$\delta E_\pi = \beta$

Since β is negative, benzene is the more stable of the pair.

A particularly simple form of Eq. 4.1 results if one fragment in the union is methyl:

$$E_\pi = 2 \sum_{s} a_{or} \beta \tag{4.2}$$

The union of methyl with pentadienyl leads to the same conclusion regarding stability as that above:

$1/\sqrt{3}$ r s $1/\sqrt{3}$

CH_3

$\delta E_\pi = 2\,(1/\sqrt{3} + 1/\sqrt{3})\,\beta$
$= 4/\sqrt{3}\ \beta$

$\delta E_\pi = 2\,(1/\sqrt{3})\,\beta$
$= 2/\sqrt{3}\ \beta$

On the surface it might appear that a more direct assessment of the DE of benzene could be obtained by considering the union of the terminal carbons in hexatriene. These two carbons are not of the same parity, and a finite bond order between them does exist. However, the determination of that bond order would require the determination of the coefficients of the MOs by the HMO method, since the PMO method does not apply to even-AHCs. The virtue of simplicity in the PMO method would be lost; hence, this is not considered a satisfactory approach.

A similar approach applied to the union of allyl and methyl yields cyclobutadiene:

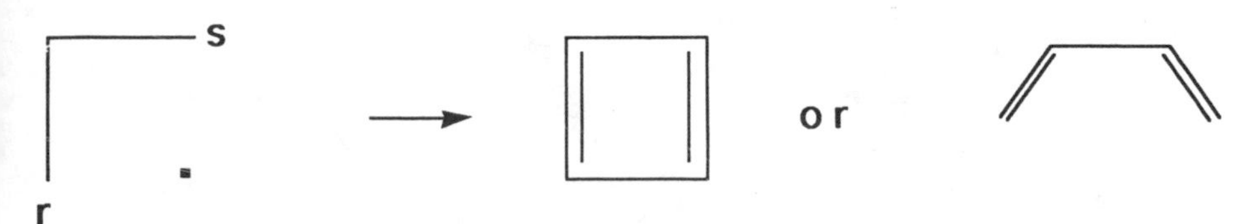

$$r = -s = 1/\sqrt{2}$$

$$\delta E_{\pi}^{\text{cyclobutadiene}} = 2(1/\sqrt{2} - 1/\sqrt{2})\beta = 0$$

$$\delta E_{\pi}^{\text{butadiene}} = 2(1/\sqrt{2})\beta = 2/\sqrt{2}\beta$$

Cyclobutadiene is less stable than its open-chain analog, which means it would actually have a negative resonance energy. Such compounds are said to be "antiaromatic." Many attempts over the years to prepare cyclobutadiene failed. However, it was finally prepared by Pettit at the University of Texas and found to have a fleeting existence in the cold and in dilute solutions. Metal coordination complexes with cyclobutadiene as a ligand have been prepared.

4.3 The Hückel Rule

During the early 1930s, Hückel produced an empirical rule from his considerations of the electronic structure of cyclic polyolefins. Those cyclics with $(4n + 2)$ π electrons were predicted to show the properties of aromatic compounds, whereas those with $4n$ π electrons would be nonaromatic. The n here is an integer, i.e., rings with 2, 6, 10, etc. electrons are to be aromatic. This stood in constrast to HMO results where all cyclic polyolefins (other than cyclobutadiene, DE = 0) were found to have appreciable DEs. Subsequent experimental work has supported the validity of the Hückel rule.

The PMO method can be shown to be in complete accord with the Hückel rule and, perhaps, can be said to provide, in theory, a real basis for its operation. Consider a linear conjugated radical of $2n - 1$ carbons. Single union with methyl will yield the $2n$ polyene, whereas double union gives the cyclic $2n$ annulene. The NBMO coefficients for the linear odd-AHC vary in a systematic fashion, i.e.,

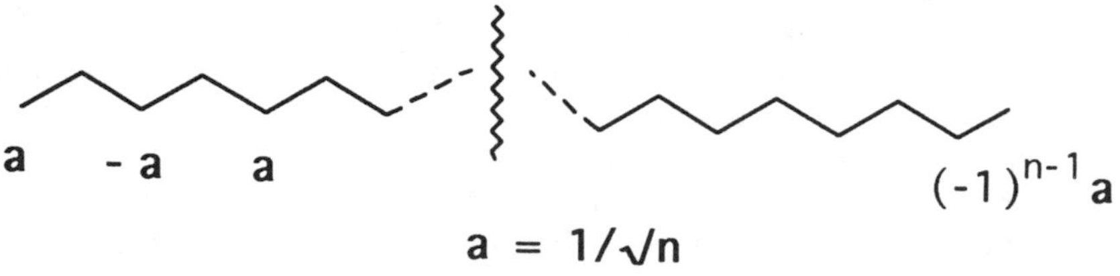

a -a a $(-1)^{n-1}$a

$a = 1/\sqrt{n}$

For the linear polyene,

$$\delta E_{\pi} = 2a\beta$$

For the annulene,

$$\partial E_{\pi} = 2(a + (-a)^{n-1})\beta \qquad (4.3)$$

Now for the cyclic system,

$$\mathrm{DE} = \frac{2(-1)^n}{\sqrt{n}}\,\beta \tag{4.4}$$

When n is odd, the annulene is more stable than the polyene (DE is positive), but when n is even, the reverse is true, and the system is antiaromatic.

In its initial form, the Hückel rule was meant to apply only to monocyclic systems, though one often finds misapplications to polycyclic compounds. Using the PMO method, it is possible to systematically apply the Hückel rule to a polycyclic compound a ring at a time. The argument, briefly, is as follows. If one denotes three contiguous atoms, r, s, and t, in one ring of a polycyclic polyene, then it can be shown that the signs of a_{or} and a_{ot} will be the same if the ring contains $(4n + 2)$ atoms. Thus, the union energy to atom s will be positive and the ring will be aromatic. If the ring contains $4n$ atoms, the signs of the coefficients will be opposite. A $4n$ ring will be antiaromatic.

Biphenylene serves as an interesting and convenient system to illustrate these points. The HMO prediction for biphenylene is a larger DE (4.51β) than for biphenyl (4.38β):

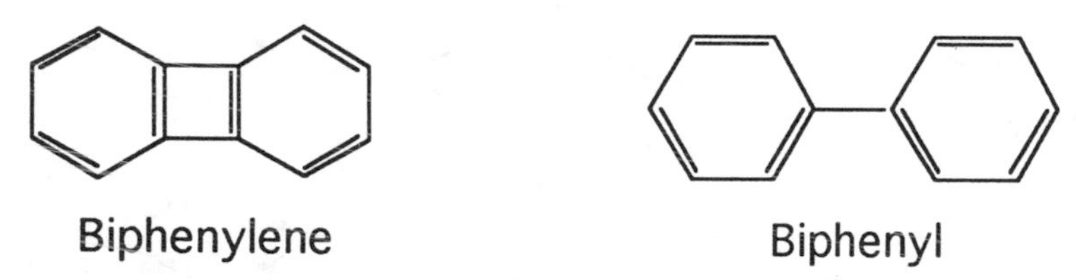

According to PMO theory, the union of two even-AHCs will, to a first order, not produce a further reduction in E_π. The bond of union between the two rings should be an essential single bond, and the DE is simply that of the two parts.

Biphenylene, biphenyl, and the open chain analog 1-phenyl-1,3,5-hexatriene may be constructed as follows:

δE_π

- a
2 a
- 2 a
a
- a ·
2 a

6 a β
8 a β
4 a β

The π energy of biphenylene is less than that of biphenyl. The two terminal six-membered rings are aromatic, while the central ring is antiaromatic. Self-consistent field calculations agree with the PMO result, as opposed to the simple HMO result.

4.4 Nonalternant Hydrocarbons

Nonalternant hydrocarbons can be handled in much the same way as benzene when one has simple structures such as fulvene. However, with more complex non-AHCs, one must exercise certain cautions. Again it is convenient to subdivide the class into even or odd, depending on whether or not there is an even number of conjugated atoms.

Azulene is an even non-AHC with two odd-numbered rings. An instructive series of compounds, including azulene, can be formed from nonatetraenyl and methyl according to the scheme below. The values shown are δE_π.

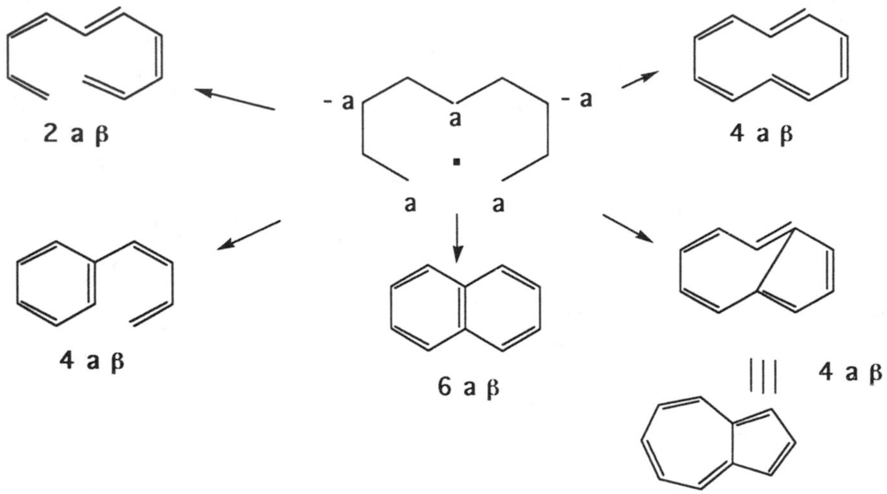

As can be seen, azulene has an appreciably lower DE than naphthalene. Furthermore, the DE is the same as that of the monocyclic polyene decapentaene, a $(4n + 2)$ π system. The bond fusing the two rings must be an essential single bond, and bond length measurements indicate this to be so.

In order to deal with odd-AHCs, it is necessary to extend the rules above. Thus, the theorem regarding bond orders between nonadjacent atoms in the same molecule must be expanded to cover the fact that bond orders in odd-AHC radicals are zero also for the union of atoms of like parity. The internal union of an allyl radical or pentadienyl radical to give the cyclopropenyl or cyclopentadienyl radicals, respectively, will result in no π energy change, since the bond orders between the two atoms uniting are zero.

When the cyclic odd-AHC is present as a cation or anion, however, there will be a first-order π energy effect when compared with the open-chain analog. It is this open-chain analog which serves as the comparison compound in making statements about aromatic or antiaromatic character. With this understanding, one may now write for the anion

$$\delta E_\pi = 2p_{rs}\beta = 2a_{or}a_{os}\beta$$

and for the cation

$$\delta E_\pi = 2p_{rs}\beta = -2a_{or}a_{os}\beta$$

As an example, we may use the pentadienyl system. Since β is negative, fusion to the anion results in a stabilizing change in π energy.

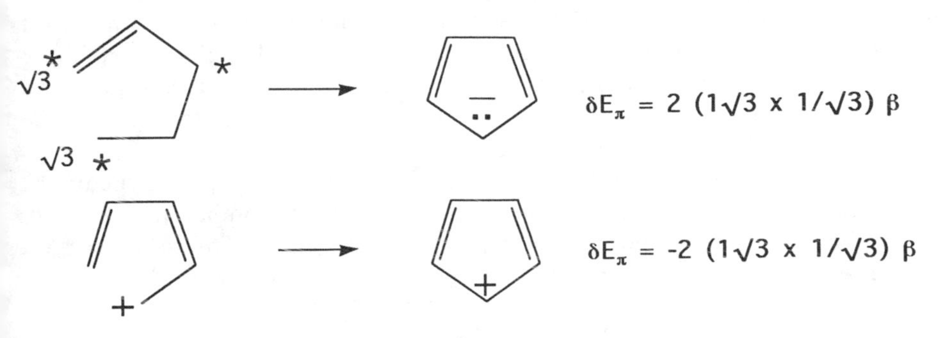

From the equations above, it is clear that systems in which the coefficients have the same sign will be aromatic, whereas those in which the signs are different will show the cation as being the aromatic form. The relative signs of the coefficients can be determined by inspection of the open-chain form or from the cyclic product. If the new ring has $(4n + 1)$ atoms, the coefficients will be alike. Rings with $(4n + 3)$ atoms will have opposite signs for the open-chain coefficients. So it follows that $(4n + 1)$ rings will have aromatic anions but antiaromatic cations, and the reverse will be true for $(4n + 3)$ systems. In either event, the aromatic system will have $(4n + 2)$ π electrons.

4.5 Resonance Theory and MO-PMO Concepts

As implied earlier, the beauty of the theory of resonance, as conventionally taught, is in its graphical nature, which allows the chemist to "think" on his feet. The facts that cyclobutadiene may be written with two Kekule structures, or that the relative stabilities for biphenyl and biphenylene are reversed from their structure counts of four and five, are considered as aberrations that one must learn to live with. The best that one can do with resonance theory, in any event, is to make qualitative statements.

Initially resonance theory was put forward as an extrapolation of the more quantitative valence bond theory. Since both involve the examination of structures written according to the rules of the electron pair bond, this implication is still often given in texts on organic chemistry. Over the years, however, a growing literature has established the close conceptual connection between resonance theory and MO theory.

We have just gone through the argument that for polyolefins such as butadiene, where only one structure can be written, MO theory, whether in the SCF or PMO approximations, predicts very little stabilization due to electron delocalization. In contrast, benzene, with its two Kekule forms, shows quite different physical and chemical properties and an appreciable DE. These results can be generalized to other linear and cyclic polyenes.

It has been known for some three decades that many of the anomalies of resonance theory can be resolved by virtue of the fact that it is possible to attribute signs to the various resonance-contributing forms. Forms with differing signs have opposite effects on stability. Thus, although one may draw more resonance forms for one molecule than for another, statements about the relative stabilities of the molecules are not warranted unless the signs of the various contributors are taken into account.

Consider first one of the Kekule structures for benzyl. Star the structure as an odd-AHC in such a way that the starred positions outnumber the unstarred. Number the unstarred positions in any order from 1 to n. The starred set are numbered from $n + 1$ to $2n + 1$.

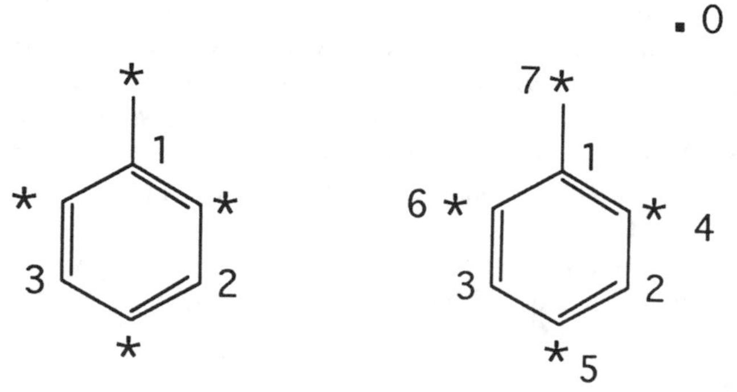

Now form pairs of one unstarred and one starred atom by pairing the doubly bound atoms of the particular structure in question. For this purpose, the odd atom may be considered to be paired with a fictitious unstarred atom 0. The bonding arrangement for the above example is represented by the set (0 7)(1 4)(2 5)(3 6). Now arrange the pairs in increasing order of the unstarred atoms 0 to n. By pairwise exchanges, rearrange the numbers $n = 1$ to

$2n + 1$ (the second number of each pair) in ascending sequence. The above structure is defined as even (plus) or odd (minus) according to whether an even or an odd number of permutations of atoms is necessary.

$$(0\ 7)(1\ 4)(2\ 5)(3\ 6) \rightarrow (0\ 4)(1\ 7)(2\ 5)(3\ 6) \rightarrow (0\ 4)(1\ 5)(2\ 7)(3\ 6) \rightarrow (0\ 4)(1\ 5)(2\ 6)(3\ 7)$$

Since three permutations are required, the sign of the structure written above is defined as negative.

It is possible to show that the other possible Kekule structure is also negative. In comparing different structures, it is important that the original numbering system be maintained. Similar treatments in which the odd atom is in the ring lead to the following results:

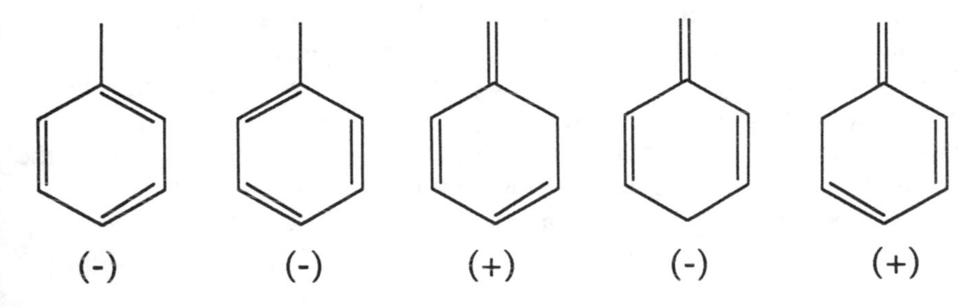

This specific case illustrates a more general point, namely, that structures which differ only in the orientation of one bond (such as 1 and 3 above) are of opposite sign. Proceeding through an argument which we shall not reproduce, it is possible to show that for an odd-AHC the number of classical structures of the same sign with the odd atom in a given position is equal to the unnormalized NMBO coefficient at that position. Thus,

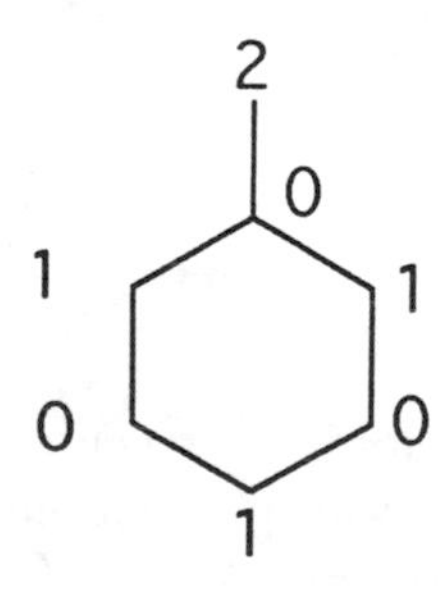

We know the odd electron in the benzyl radical has its highest probability density at those active sites where the amplitude of the NBMO coefficients is greatest. One sees here a direct link between resonance theory and MO results. Obviously, the same reasoning may be carried over to the case of the benzyl cation or anion.

Further generalizations for the odd-AHC case may be cited. Consider pentadienyl and α-methylnaphthyl along with benzyl from above:

The total number of classical structures for each case is given by the sum of the absolute values of the coefficients. Their unions with methyl yield, for example, styrene, hexatriene, and α-vinylnaphthalene. These compounds have a total of two, one, and three classical structures, respectively, just the value of the coefficient at the point of union. The same conclusion holds for the removal of the odd atom from the conjugated system; benzene, butadiene, and naphthalene also show two, one, and three structures. These observations point the way to how we might handle resonance counting in even-AHC systems.

The important point is not the absolute sign associated with a given group of structures, but whether the signs are the same or different throughout the group.

Let us consider cyclobutadiene. We could do it as above, using methylenecyclobutadiene. Instead we will treat the two Kekule forms directly:

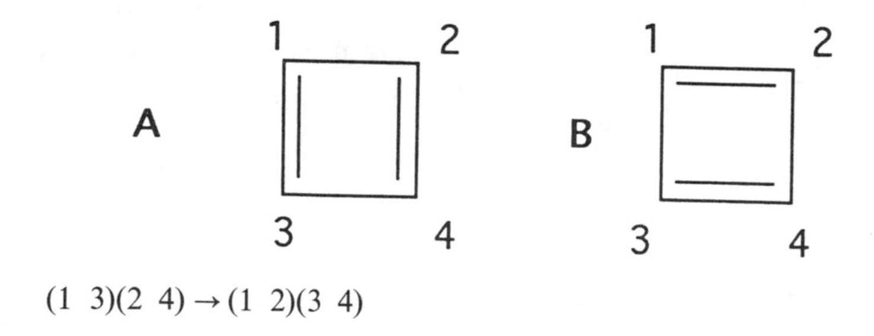

There is one permutation. Therefore, the structure A is opposite in sign from B.

In our earlier HMO treatment of cyclobutadiene, two NBMOs were discovered in what was clearly an even-AHC. It can be shown that in even-AHCs where half the structures are of one sign and half are of the other sign, similar NMBOs, equal in number to the structures written, will be found. These orbitals are termed "supernumerary" orbitals. Such molecules will exhibit diradical ground states if planar (cyclobutadiene) or will twist into the nonplanar puckered configuration (e.g., cyclooctatetraene). Benzenoid structures having $(4n + 2)$ membered rings will have classical structures all of the same sign and with no supernumerary orbitals.

The biphenylene case may be used to make another point. The resonance structures are given here.

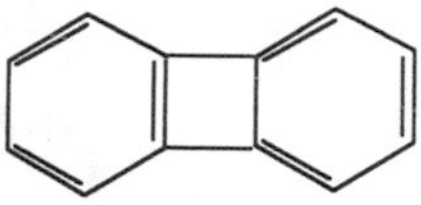

A 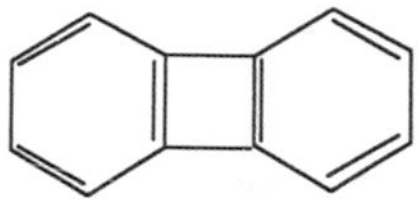 B 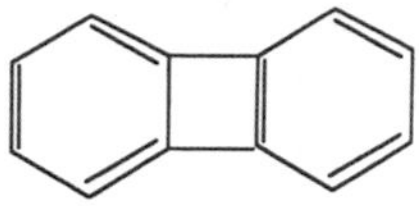 C

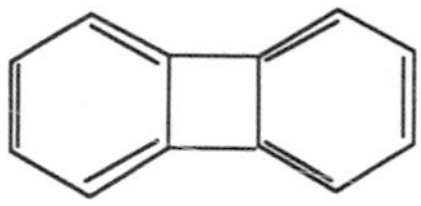

D 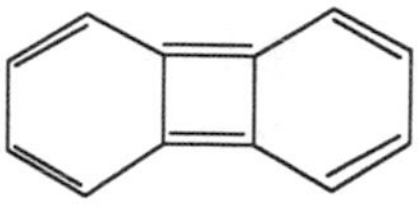 E

The signs of each structure could be determined, as for benzyl, by adding an α-methylene group, or the relative sign of each structure with reference to A may be ascertained as previously for cyclobutadiene. In either event, one concludes that A, B, C, and D are of one sign and E is of the opposite sign. The algebraic structure count (ASC) is the difference between the number of structures of opposite sign. For biphenylene the ASC is 3. The structure count (SC) is five. In biphenyl all structures are of the same sign, since both rings are benzenoid, and thus the ASC and SC are the same. The resonance theory with signs correctly assigned peredicts biphenyl (ASC 4) to be more stable than biphenylene (ASC 3). The result, of course, is in accord with PMO conclusions. Another generalization may be offered to handle even-AHCs with one or more $4n$ rings, provided they are not fused to each other. The ASC in such systems is that number remaining after all structures having $2n$ double bonds in the $4n$ ring(s) have been deleted. For instance, I has been found to be more stable than II:

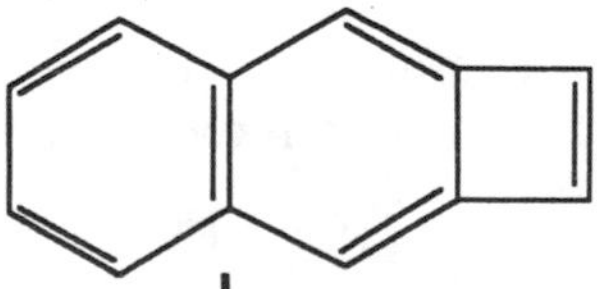

I
ASC 2

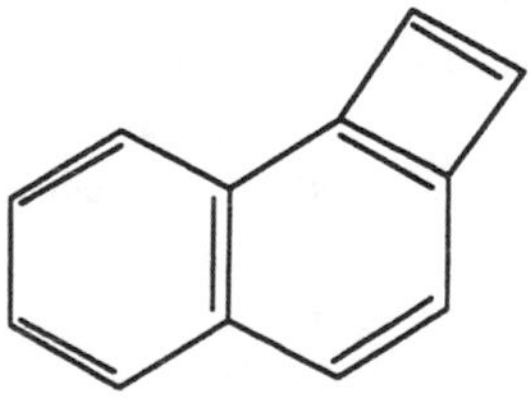

II
ASC 1

Yet another technique exists for determining values of SC and ASC. For even-AHCs, one may delete an atom generating an odd system. With fused-ring systems, one must choose an atom at a fusion point. The NBMO coefficients for the odd-AHC are then determined. The sum of the absolute values of the unnormalized coefficients about the deletion point equals the SC, whereas the absolute value of the sum equals the ASC.

SC 4, ASC 2

SC 5, ASC 1

Interestingly, the atom deletion technique works quite well with non-AHCs. Application to azulene (as well as for naphthalene for comparison purposes) is shown below.

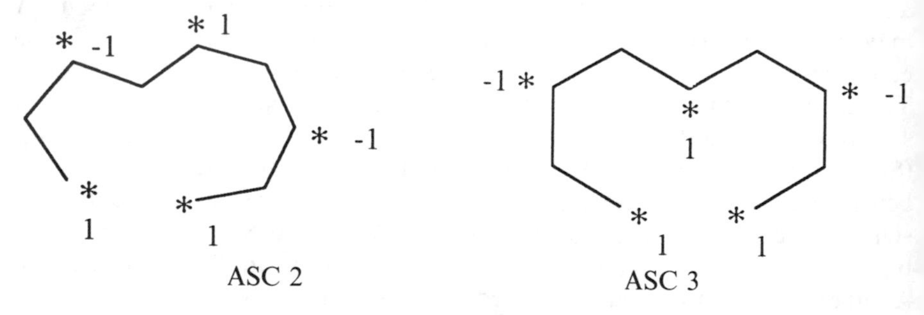

ASC 2

ASC 3

According to our earlier PMO treatment, the multiple union of two odd-AHCs to form an even-AHC is accompanied by an energy change

$$\delta E_\pi = 2(a_{or}b_{os} + a_{oq}b_{ot})\beta \tag{4.5}$$

where a_{or}, a_{oq} and b_{os}, b_{ot} are the appropriate NBMO coefficients in the two fragments. Now if the two fragments are benzenoid, the absolute values of the unnormalized coefficients expressed as whole numbers are the SCs with the odd atoms located at r and s in one and at q and t in the other. Equation 6.5 may be rewritten as

$$\delta E_\pi = 2K(n_r m_s + n_q m_t)\beta \tag{4.6}$$

where the n and m values are the structure counts. The quantity in brackets in Eq. 6.6 is the total number of classical structures that can be written for the united molecule.

Using benzyl as an example, one may form each of the following:

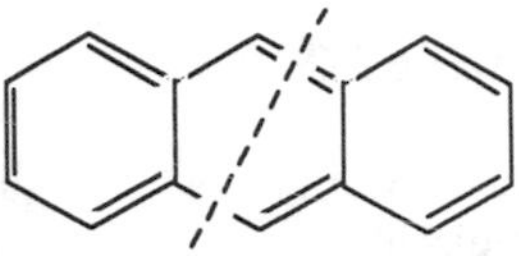

SC 4
(2x1 + 2x1)

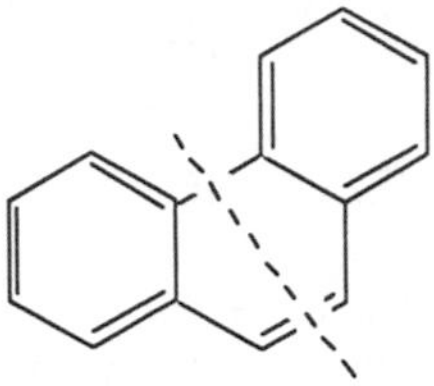

SC 5
(2x2 + 1x1)

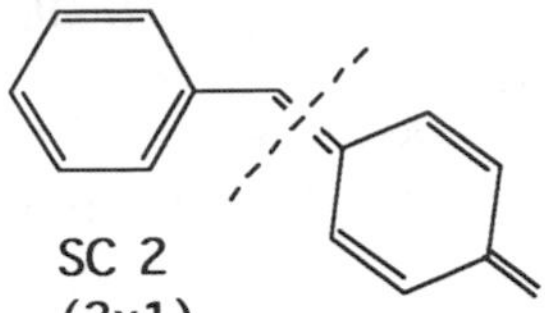

SC 2
(2x1)

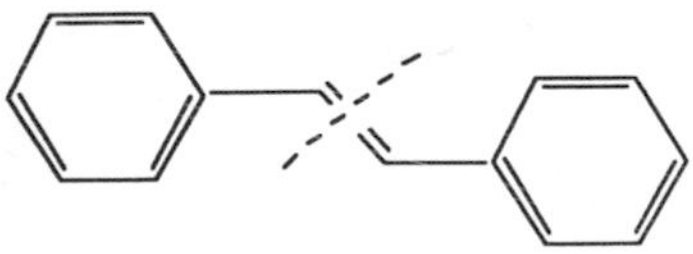

SC 4
(2x2)

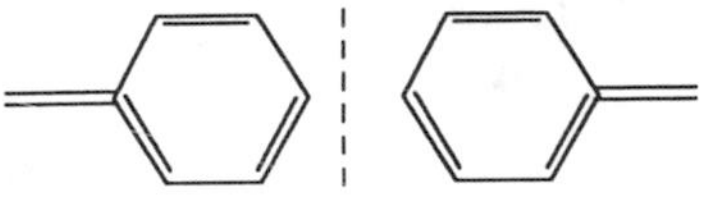

SC 1
(1x1)

The union energy for phenanthrene is greater than for anthracene, a result in keeping with experimental resonance energies and the generalization that angular fused systems are more stable than their linear counterparts. The relative stabilities of the quinoid structure are also as expected.

Problems

1. Using the bond energies of Hess and Schaad, calcualte E_π for localized fulvene. From the HMO value of E_π (Chapter 4, problem 2), calculate the DE for fulvene. Contrast this with the conventional DE calculated in the earlier problem.

2. Using the PMO method, calculate the DE for fulvene by the union of methyl with pentadienyl. (Note: the comparison compound is hexatriene.)

3. Calculate the DEs for biphenyl, biphenylene, naphthalene, and azulene using the localized bond energies of Hess and Schaad. The HMO values for E_π are 16.38, 16.51, 13.68, and 13.36β, respectively. Compare these DEs with those predicted on the usual Hückel basis. Also compare the results with PMO predictions.

4. Calculate δE_π for 1,3,5,7-octatetraene, cyclooctatetraene, styrene, benzocyclobutadiene (below), and pentalene (below). Use the union of heptadienyl and methyl for one series of calculations. Then repeat using the union of pentadienyl and allyl. (Warning: Application of the PMO method requires the application of common sense as well as the rules. Bonds will tend to form between active positions in preference to inactive positions.)

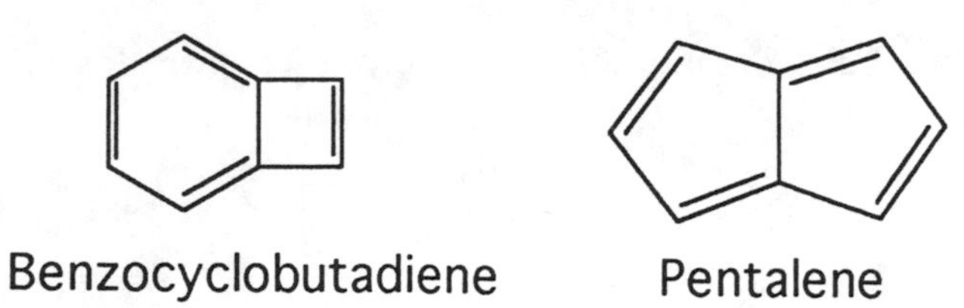

5. Determine the SC and the ASC for each of the molecules below. Are the results consistent with the fact that only IV and V are known molecules?

I II III

IV V

CHAPTER

5

Hückel and FMO Theories Applied to Chemical Reactivity

5.1 The Nature of the Problem

The prediction of reaction products, rates of reaction, and reaction equilibria is one of the prime goals of quantum chemistry. However, at present it is not possible to calculate reaction rates from first principles except for perhaps the simplest molecular systems. What is required, of course, is a knowledge of the free energy of reaction for equilibria and the free energy of activation for rate processes. The latter requires a knowledge of the transition state, which, in turn, implies information on the whole potential energy surface for the reaction. In later chapters we will examine how the latest computational programs conduct searches on that surface. As will be seen, the breakdown of free energy into enthalpy and entropy offers direct help in the solution of the problem, since entropy calculations, which require a knowledge of vibrational and rotational frequencies for reactants and transition states, are becoming increasingly more accurate and available.

In this chapter attention will be focused on asking questions about relative reactivities. How reactive are various positions within the same molecule towards a given reagent? What differences in reactivity occur among a series of related molecules undergoing the same reaction? Inherent in any answer to these questions is the assumption either that all entropy changes are the same in the systems under examination or that there is some systematic variation in entropy that is compensated for in the enthalpy term. Experimentally it has been demonstrated that entropies of reaction and activation entropies often vary widely among the members of a reaction series. Happily there are many cases known in which linear plots of entropy versus enthalpy are found. Organic chemists have a surprising list of reactions demonstrating logical correlations between reaction rates or equilibria and variations in molecular structure.[1]

Finally, if we are to calculate reaction rates, we need to have energy differences between the ground state and the transition state for the reaction. This implies a knowledge of the structure of the transition state, which is sometimes hard to come by. In 1955 George Hammond gave formal expression to a rule of thumb found in many works dealing with rate phenomena. Hammond's postulate says, in essence, that a transition state with a free energy close to that of a reactive intermediate will have structural features closely resembling the intermediate. Since we have information about the reactive intermediates in many reactions, it follows that we often have knowledge of the nature of the transition states too.

5.2 Electrophilic Substitution

Substitution reactions in aromatic systems involving attack by an electron-seeking reagent have been exhaustively studied over the years. Since much of the pertinent evidence regarding the mechanism can be found in organic chemistry texts, no review of the facts will be given here.

It is considered that attack on the π system by the electrophile Y^+ leads to the formation of a reactive intermediate with a pentadienyl cation structure. Collapse back to a fully aromatic product follows:

H Y

Y

+ Y^+ → → + H^+

By Hammond's postulate, the transition state for forming the high-energy intermediate should look very much like the intermediate. If so, then the main energy change in reaching this transition state will be that of localizing two electrons from the π system in the developing bond to Y. The loss of delocalization energy in going to the pentadienyl cation will measure this.

Wheland some years ago suggested the use of the localization energy as a useful reactivity index:

$$\Delta E = E_{\mathrm{OA}} - E_{\mathrm{EA}} \tag{5.1}$$

Here E_{OA} is the π energy for an odd-AHC system such as pentadienyl, whereas E_{EA} is the π energy for an even-AHC system of one more atom, such as benzene. From HMO calculations, these values are $4\alpha + 5.46\beta$ and $6\alpha + 8.00\beta$, respectively. The value of the localization energy is 2.54 in units of $-\beta$ to keep the value positive. The α terms cancel, since each localized π electron contributes α to the energy sum.

The above calculation for substitution in benzene requires the complete solution of two secular determinant. The PMO method offers a far quicker solution. We saw earlier (p. 55) that the energy change in going from pentadienyl to benzene could be approximated by the union with methyl. This is precisely the energy change we require, but with the sign reversed. Thus,

$$\Delta E_{\text{loc}} = -2 \sum p_r \beta = -2 \sum c_{or} \beta \tag{5.2}$$

where the C_{or} values are the NBMO coefficients of the positions being joined to the one-carbon fragment. Applied to benzene:

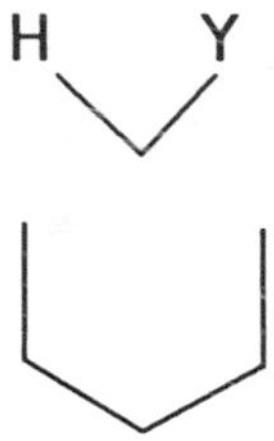

H Y

$$\Delta E_{\text{loc}} = -2(1/\sqrt{3} + 1/\sqrt{3})\beta$$
$$= -2.31\beta$$

Dewar has defined the term $2\Sigma c_{or}$ as N_t, the reactivity number. In general, it may be said that values of N_t are smaller than HMO localization energies, but they parallel the latter and reflect reactivities as well.

Electrophilic attack on naphthalene can occur at either the α or the β position:

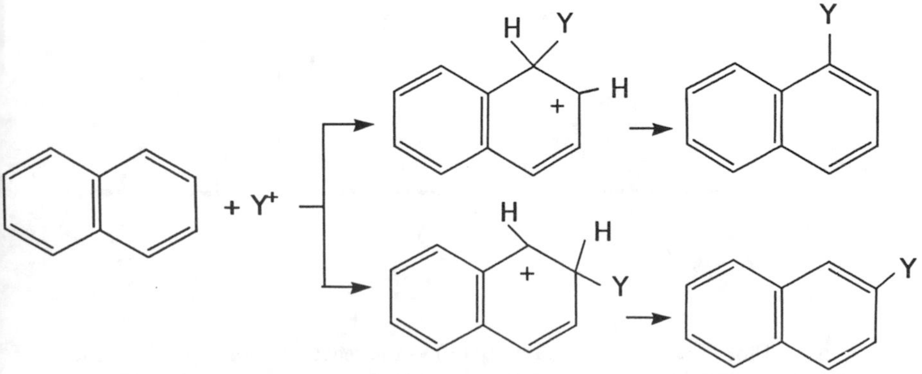

+ Y⁺ H Y + H → Y

H H + Y → Y

The energies of the two transition states may be approximated by the two systems below:

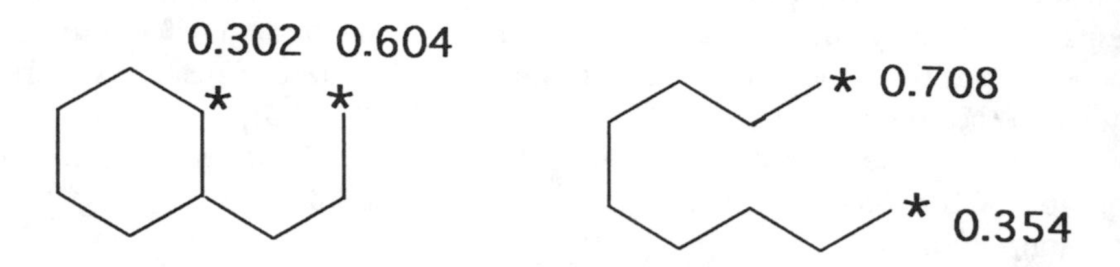

The numbers are the NBMO coefficients. The N_t for α substitution is $2(0.302 + 0.604) = 1.81$, whereas that for β substitution is $2(0.708 + 0.354) = 2.12$. The lower reactivity number for α substitution is in agreement with the experimental observation that α products are kinetically first formed from naphthalene. For reversible reactions such as sulfonation, one may actually isolate the thermodynamically more stable β isomer.

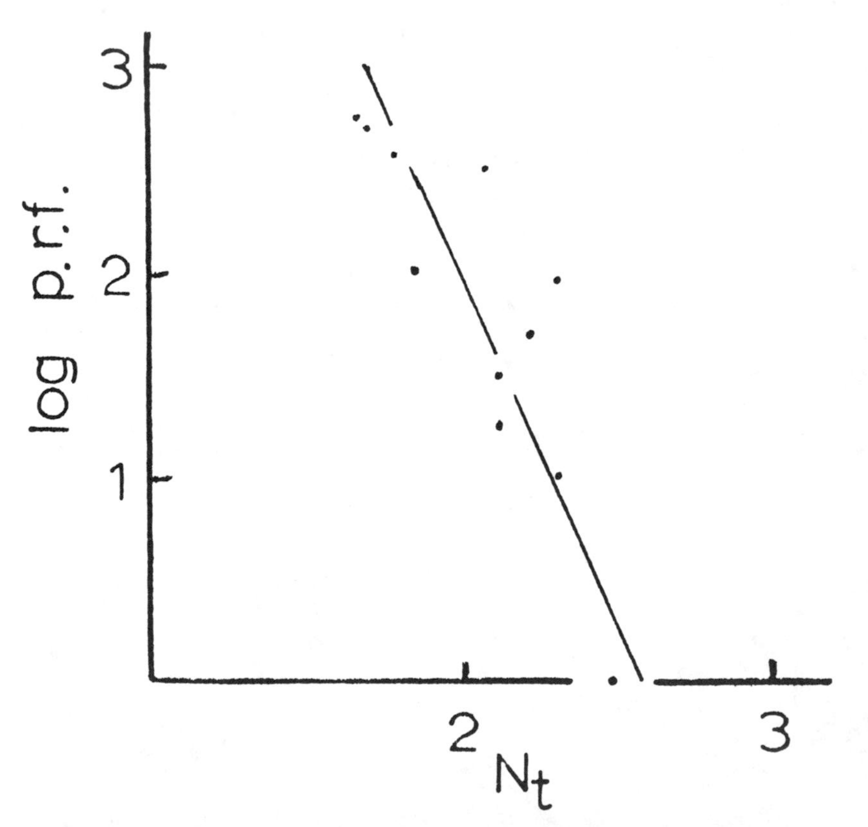

Figure 5.1. Plot of partial rate factors (p.r.f.) vs the reactivity number (N_t) for a series of polynuclear aromatic hydrocarbons undergoing nitration in acetic anhydride.

The reactivity numbers for benzene, naphthalene, and anthracene are given below. Not only are the correct positions of substitution predicted for the latter two, but the relative ease of substitution in the series is correct.

As a case in point, Figure 5.1 shows the plot of the log of partial rate factors for nitration for series of aromatic hydrocarbons against N_t.

5.3 Heteroatom Systems

As demonstrated in Chapter 3, heterocyclic molecules may be considered as perturbed versions of their carbocyclic analogs. In the case of reaction rate calculations, we must know not only how the heteroatom perturbs the reactant but also how it perturbs the transition state. Let the change in π energy on introducing the heteroatom into the even-AHC be δE_{EA}:

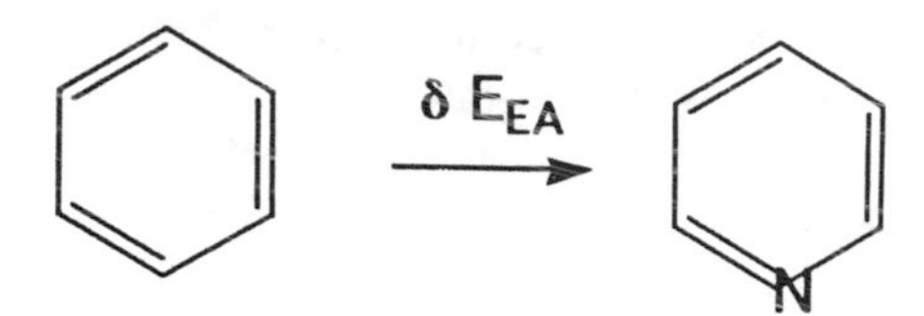

For the intermediate (an odd-AHC), let the same process be given by δE_{OA}:

The localization energy for the electrophilic substitutions energy for pyridine may now be approximated by

$$\Delta E_{loc} = E_{OA} - E_{EA} + \delta E_{OA} - \delta E_{EA} \tag{5.3}$$

Maintaining the assumption that $\beta_{CN} = \beta$ from p. 27 and remembering Eq. 3.21, we can write the energy change for any orbital (u) upon introducing a heteroatom as

$$\delta E_u = c_{ui}^2 \delta\alpha_i \tag{5.4}$$

where again C_{ui} is the u-th MO coefficient of the atom i. The total change of energy (δE) requires multiplying δE_u by the number of electrons (N) in the orbital and then summing over all occupied orbitals:

$$\delta E = \sum_u^{\text{occ}} N_u c_{ui}^2 \delta\alpha_u = \text{ED}_1\, \delta\alpha_1 \tag{5.5}$$

The term $\sum_u^{\text{occ}} N_u c_{ui}^2$ is just our earlier definition of total electron density. Since all carbons in a neutral AHC have electron densities of unity, Eq. 5.5 reduces to

$$\delta E_{\text{EA}} = \delta\alpha_i \tag{5.6}$$

for the conversion of benzene to pyridine.

The odd-AHC fragment above bears a charge. Consequently, the change in π energy is dependent upon orbital occupancy. For odd-AHCs, only radicals have unit electron densities at each position. A cation has one less electron than the radical, that electron being removed from the singly occupied NBMO. The electron density in the cation is then simply

$$2 \sum_{u=1} c_{ui}^2 = 1 - c_{oi}^2 \tag{5.7}$$

and therefore

$$\delta E_{\text{OA}} = (1 - c_{oi}^2)\delta\alpha_i \tag{5.8}$$

By virtue of Eq. 5.8, we can now get to the δE_{OA} term of Eq. 5.3 with only a knowledge of the NBMO coefficients.

Substituting these results into Eq. 5.3 allows us now to write for electrophilic substitution in the heterocycle

$$\Delta E_{\text{loc}}^{+} = \Delta E_{\text{loc}} - c_{oi}^2 \delta\alpha_i = -N_t - c_{oi}^2 \delta\alpha_i \tag{5.9}$$

where ΔE_{loc} is the localization energy for the carbocyclic analog.

It may be well to mention here before proceeding that similar approaches can be taken to free radical and nucleophilic substitution reactions. The results for these cases have been shown to be

$$\Delta E_{\text{loc}}^{\text{rad}} = \Delta E_{\text{loc}} \quad \text{(free radical)} \tag{5.10}$$

$$\Delta E_{\text{loc}}^{-} = \Delta E_{\text{loc}} + c_{oi}^2 \delta\alpha_i \quad \text{(nucleophilic)} \tag{5.11}$$

Let us now apply Eq. 5.9 to pyridine. The three possible intermediates representing their respective transition states are given below, as are the NBMO coefficients for pentadienyl:

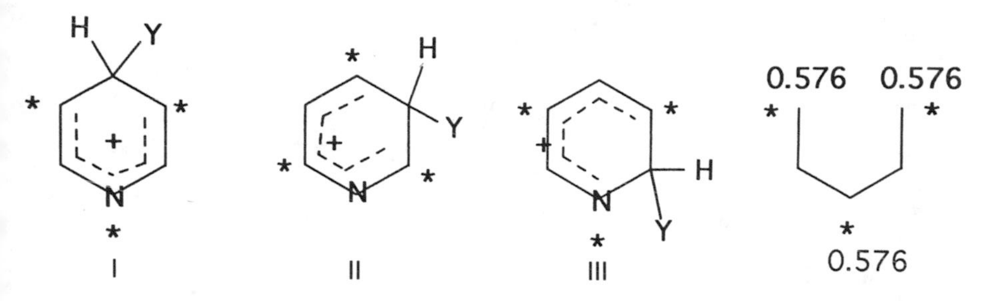

Note that in I and III the heteroatom appears at an active position in the pentadienyl fragment. Thus, the energies of these will be perturbed by $c_{oi}^2 \delta\alpha_i$. Since for nitrogen this term is usually approximated by 0.5β, we may write for each of the three cases above

$$\text{(I)}\ \Delta E_{\text{loc}}^{+} = -2.31\beta - (0.576)^2 \times 0.5\beta = -2.78\beta$$

$$\text{(II)}\ \Delta E_{\text{loc}}^{+} = -2.31\beta - (0)^2 \times 0.5\beta = -2.31\beta$$

$$\text{(III)}\ \Delta E_{\text{loc}}^{+} = -2.31\beta - (0.576)^2 \times 0.5\beta = -2.78\beta$$

The lowest value for the localization energy for electrophilic attack in pyridine is to be found in II, and, of course, this is where substitution does occur. However, the localization energy calculated here is the same as that for benzene itself, and this stands in marked contrast to the experimental fact that pyridine is considerably less reactive than the carbocycle. It has been speculated that the electronegativity of the nitrogen causes a general lowering of the electron density of the ring system compared to benzene, which is reflected in a lowered reactivity at the 3 position. Another possibility is that of repulsion between a positively charged electrophile and the pyridinium ion that would be present in nitration or sulfonation reactions or other substitutions in acid media.

5.4 Free Radical Substitution

Detailed rate studies of free radical substitution processes are not available. However, relative reactivities toward methyl and trichloromethyl radicals have been ascertained. The best evidence supports a mechanism resembling that for electrophilic substitution. If so, then a similar relationship between reactivity

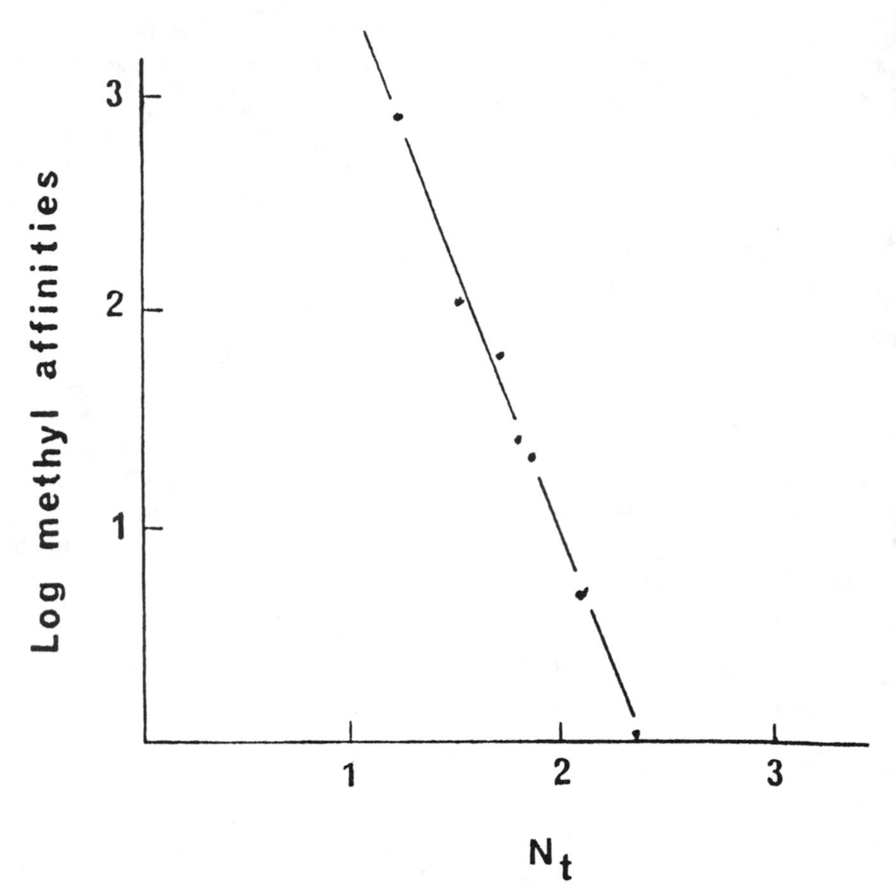

Figure 5.2. Plot of reactivity numbers (N_t) versus the methyl affinities for a series of polynuclear aromatic hydrocarbons.

numbers and the log of the reactivity index (called here the methyl affinity) should be noted. Since no product distributions have been determined for the polynuclear aromatic hydrocarbons, the plot shown in Figure 5.2 uses the reactivity number at the most reactive position only.

5.5 Nucleophilic Substitution

Nucleophilic substitution reactions on substituted benzenes, in general, occur only with considerable difficulty. However, when a nitrogen atom is inserted in the ring, such substitutions may take place readily. The situation is covered by Eq. 5.11, the delocalization energy being reduced by the $c_{oi}^2 \delta\alpha_i$ term. This term can only be important if the heteroatom is located at an active position in the intermediate. This, of course, means that the nitrogen must be *ortho* or *para* to the position of substitution. It goes without saying that molecules with multiple heteroatoms in strategic positions react even more readily.

5.6 Paralocalization Energy

In the next chapter we will consider in some detail a variety of multicenter reactions. These are characterized by the fact that several atomic centers in one or more molecules come together in a concerted breaking of old bonds and making of new ones.

One such reaction is the Diels-Alder reaction, and it is appropriate that we anticipate the subsequent discussion while considering localization energies. The reaction of anthracene with maleic anhydride is often cited as an example of the Diels-Alder reaction:

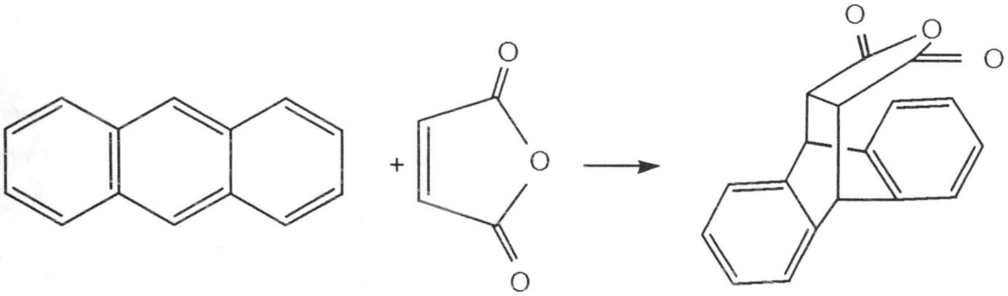

One may well wonder why the above reaction occurs exclusively at the 9,10 positions. Some years ago it was shown that such reactions correlate with the localization energies of the carbons *para* to each other in the aromatic systems. Paralocalization energies correspond approximately to the sum of the reactivity numbers for these positions (in units of $-\beta$). The values for anthracene are shown here, and not unexpectedly, the paralocalization energy is lowest for the 9,10 positions. The paralocalization energies for naphthalene (3.62) and benzene (4.62) are considerably higher than the values for anthracene, and they do not form adducts under normal conditions.

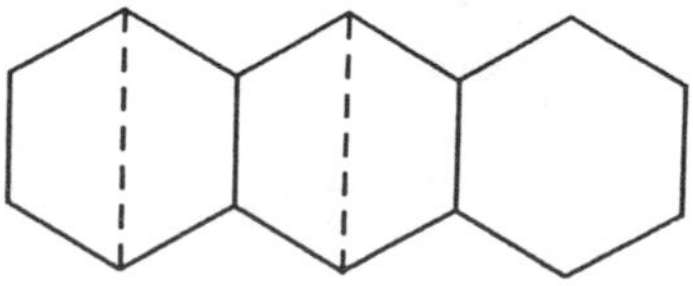

2.52

3.14

5.7 Ortholocalization Energy

Certain aromatic hydrocarbons react with osmium tetraoxide to form osmate esters, which can be hydrolyzed to diols:

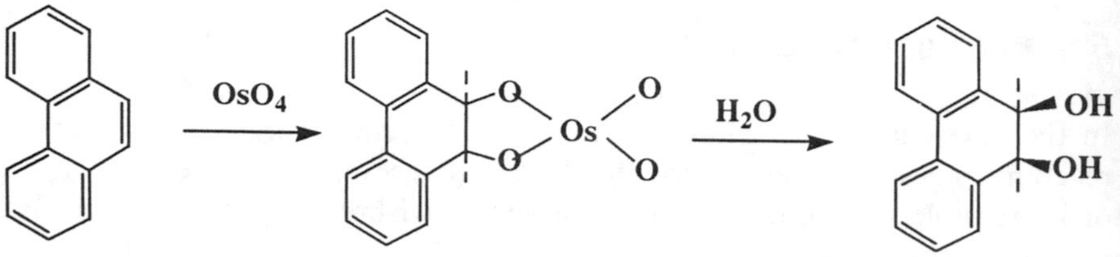

The mode of addition is *cis*, and clearly the reaction should be sensitive to the bond attacked, as a portion of the conjugated system is destroyed. The effect on E_π of the loss of these two adjacent centers could be calculated from MO theory, as in the case of the localization energy (Eq. 5.1). However, Dewar has shown that a version of Eq. 5.2 can be used to obtain approximate ortholocalization energies, i.e.,

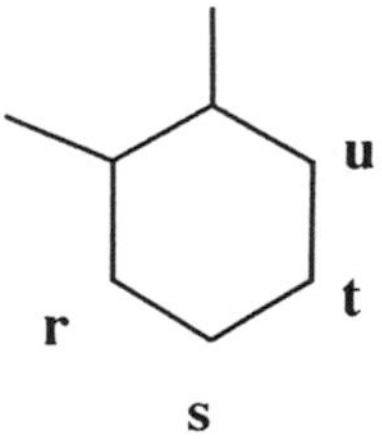

The requisite bond orders can be obtained, per Eq. 3.27, by union of the appropriate odd-AHC with methyl. As an example, one might use phenanthrene. For carbons 9 and 10, we can write

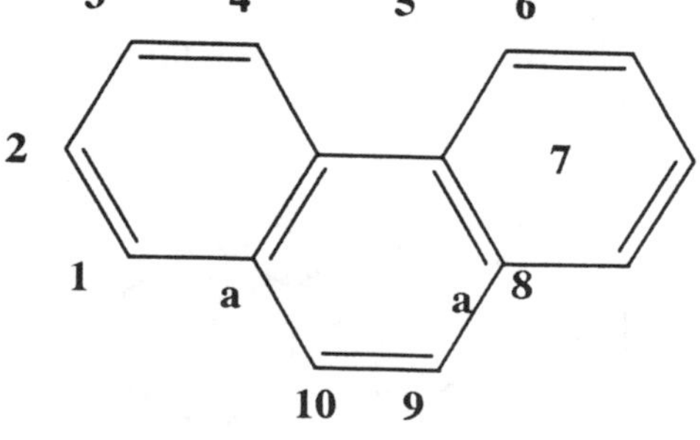

$$\Delta E_o = -2(p_{a9} + p_{9,10} + p_{a'10})$$

By symmetry, p_{a9} equals $P_{a'10}$. Then from the diagram below, it follows that

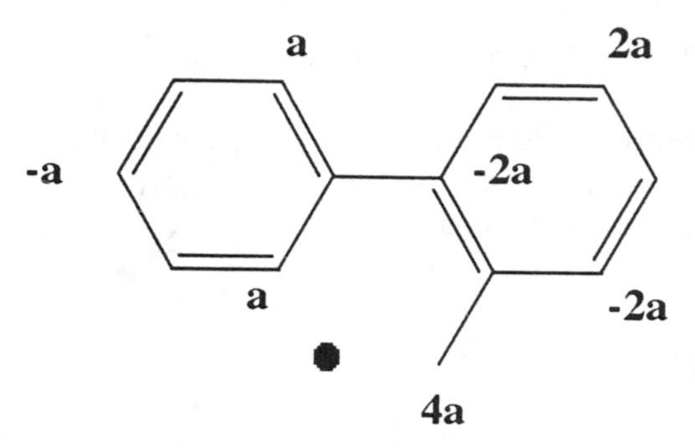

$$31a^2 = 1;\ a = 1/\sqrt{31} = 0.179$$

$$p_{a9} = p_{a'10} = 1 \times 0.179 = 0.179$$

$$p_{9,10} = 1 \times 0.716 = 0.176$$

and

$$\Delta E_0 = -2(2 \times 0.179 + 0.716)\beta = -2.15\beta$$

The ortholocalization energy is usually expressed as a positive number (i.e., in units of $-\beta$).

When the bond is not as symmetrically located, it is clear that one might calculate more than one bond order for the value between the carbons where localization is to occur. Let us consider the 1–2 bond in phenanthrene.

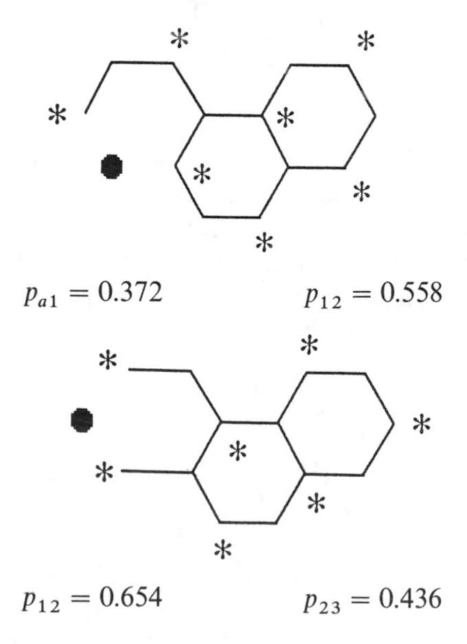

To calculate ΔE_0, the average value for p_{12} is used:

$$\Delta E_0 = -2(0.372 + \tfrac{1}{2}(0.558 + 0.654) + 0.436)\beta = -2.828\beta$$

The bond with the lowest ortholocalization energy is the 9–10 bond (2.15), and it is here that the reaction occurs. A study of reactive sites in a substantial number of aromatic hydrocarbons has shown an excellent correlation.

5.8 Carcinogenic Activity

Some years ago, the French research team of B. and A. Pullman set about studying the electronic properties of those polynuclear hydrocarbons which were known to be capable of inducing cancer. MO calculations on these substances showed regions of olefin-like bond character designated as K regions.

Subsequently, it was shown that the character of various transannular regions (L regions) was also required to characterize carcinogenic potentials. The two regions are shown here for benzanthracene (which, incidentally, is not carcinogenic):

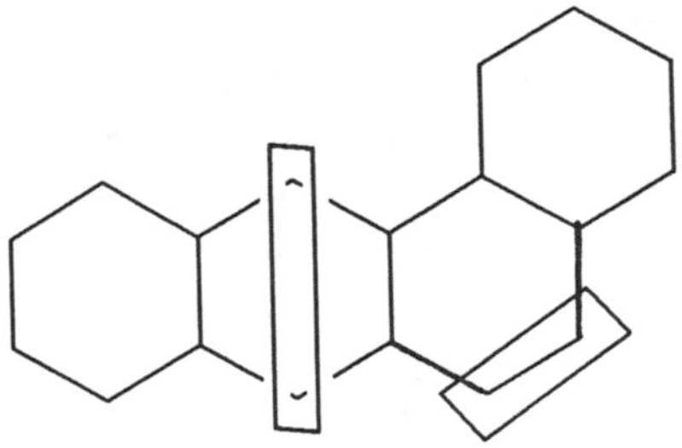

The paralocalization and ortholocalization energies approximated from PMO theory may be used for the characterization. The requirement for carcinogenic character appears to be ortholocalization energies below about 2β provided there is no region of paralocalization energy of less than about 2.9β.

Two examples of particularly potent carcinogens are

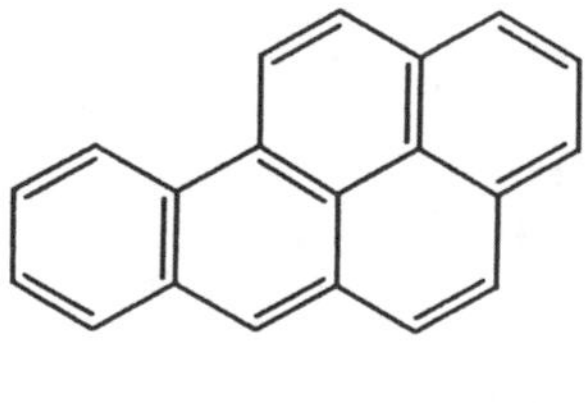

3,4-Benzopyrene

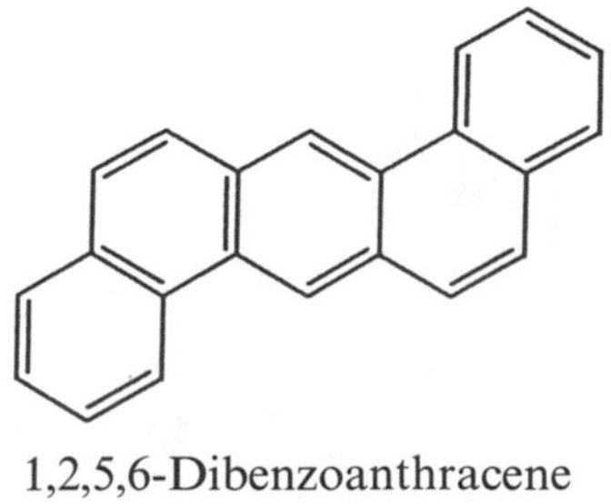

1,2,5,6-Dibenzoanthracene

5.9 FMO Theory Applied to Aromatic Substitution

Starting in the early 1950s, an important series of papers emerged from Japan authored by Kenichi Fukui and co-workers. One of the earliest of these dealt with the subject of electrophilic aromatic substitution.[2] Other aspects of this topic will be taken up in the next chapter.

The essentials of Fukui's argument are as follows. An electron-deficient reagent (say a bromonium or nitronium ion) approaches an aromatic π system. In seeking to satisfy its electrophilic character, what does the reagent first encounter? The answer, of course, is those electrons in the highest occupied molecular orbital (the HOMO). The contribution of these electrons to the π charge density is proportional to the square of the MO coefficients for each carbon, though we can make a judgment of these values by looking at the absolute values for the HOMO coefficients themselves.

The calculations were carried out on the AHCs shown below using the standard HMO method. The numbers correspond to the HOMO coefficients at that atom. The arrows point to the positions of greatest attack by an electrophilic reagent as taken from experimental data.

Although the above simple method applied well to various AHCs, difficulties were experienced when the method was applied to substituted benzenes and to aromatic heterocyclic molecules. The frontier orbitals of benzene are

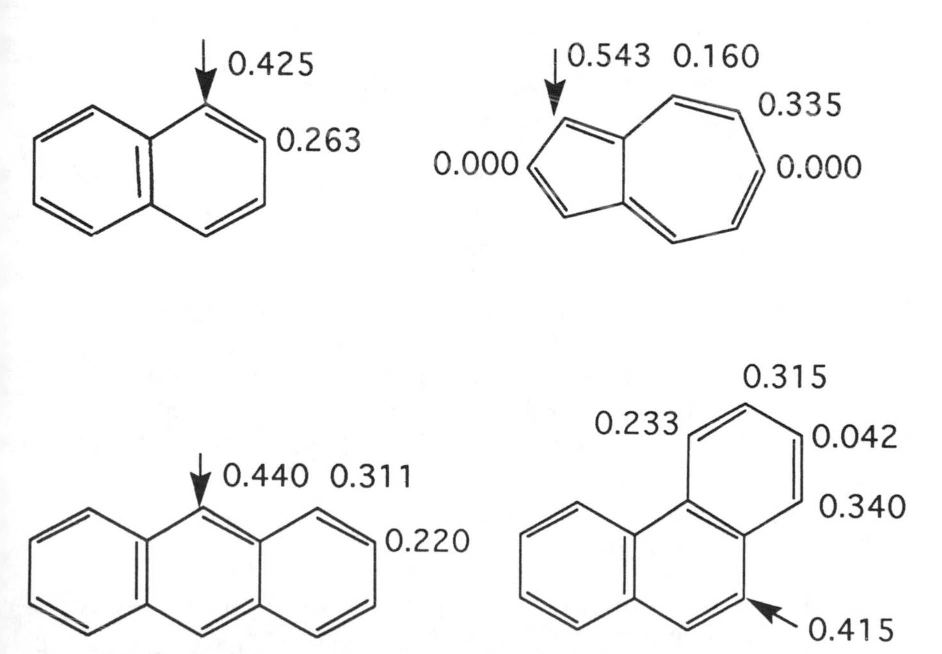

Figure 5.3. HOMO coefficients (absolute values) for some AHCs and the experimental positions of preferred electrophilic substitution.

degenerate. This degeneracy is removed by attaching a substituent to the ring or by replacing one of the ring carbons with a heteroatom such as oxygen or nitrogen. Seemingly, one must consider the orbital immediately below the HOMO, as these fall very close in energy to the HOMO itself. The following formula was devised to address this situation:

$$f_r^{(E)} = 2\,\frac{|C_r^{(1)}|^2 + |C_r^{(2)}|^2 e^{-D\Delta\lambda}}{1 + e^{-D\Delta\lambda}} \tag{5.12}$$

where $\Delta\lambda$ is the energy difference between the HOMO and the orbital immediately below it given in units of β. The C values are the coefficients of the HOMO and the adjacent orbital for atom r, and D is a constant assigned the value of 3 in most studies. Application of Eq. 5.12 is given for two cases below, where the energies of the orbitals were determined by a standard HMO calculation:

Cl
0.348
0.213
0.367

0.272
0.402
0.336
N

Orbital energies up to the HOMO in β units

1	2.609	2.107
2	1.790	1.167
3	1.000	1.000
4	0.755	

These results are consistent with the known substitution patterns of both molecules with regard to electrophilic substitution. Equation 5.12 has been applied also to certain examples of nucleophilic and free radical aromatic substitution, and the original literature should be consulted for those examples.[2] Additional applications of FMO theory will be given in Chapter 6.

Problems

1. Calculate N_t for the various carbons in phenanthrene and show that these are roughly in accord with the relative order of nitration, i.e., with benzene as 1 the values for the various positions in phenanthrene are (1) 360, (2) 92, (3) 300, (4) 79, and (9) 490.

2. Calculate the expected position(s) of electrophilic attack in quinoline and in isoquinoline.
3. Compare the ortholocalization energies of 3,4-benzopyrene with those of benzene and naphthalene, and with benzanthracene.
4. Show that resonance theory and PMO theory lead to different conclusions regarding the position of electrophilic substitution in biphenylene. The PMO method gives the correct answer in this case.
5. Account for the fact that pyridine undergoes nucleophilic substitution by amide ion to form either 2-amino or 4-aminopyridine in preference to 3-aminopyridine.
6. Predict the preferred position of attack by dienophils on phenanthrene.

References

1. See, for instance, J. E. Leffler and E. Grunwald, *Rates and Equilibria of Organic Reactions*, John Wiley, New York, 1963; reprinted by Dover Publications, New York, 1989.
2. K. Fukui, T. Yonezawa, and H. Shingu, *J. Phys. Chem.*, **20**, 722 (1952); for a more detailed account and leading references, see K. Fukui, T. Yonezawa, and C. Nagata, *ibid.*, **26**, 831 (1957).

CHAPTER

6

Pericyclic Reactions, Orbital Symmetry, and PMO and FMO Theories

As evidence regarding organic reaction mechanisms was acquired and digested during the earlier decades of this century, several types of reactions became conspicuous by virtue of the absence of any indications of reactive intermediates. For the most part, these cases had in common their rather exact stereospecificity and the absence of any catalysts other than the input of energy in the form of heat and light.

To some it appeared that these reactions might involve cyclic, neutral transition states, which were depicted as being formed by electrons moving along the paths of curved arrows, i.e.,

$$C_6H_5-O-CH_2-CH=CH_2 \longrightarrow o\text{-}HO-C_6H_4-CH=CH_2 \quad (6.1)$$

Named reactions such as the Diels-Alder reaction, the Cope rearrangement, and the Claisen rearrangement fell in this category.

Recently, it has become customary to use more exacting definitions descriptive of the reaction type. The three current classifications are exemplified by the following:

Electrocyclic reactions

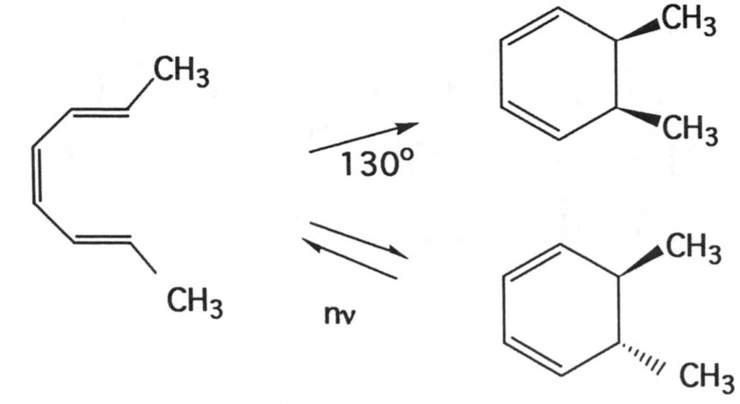

(6.2)

Cycloaddition reactions

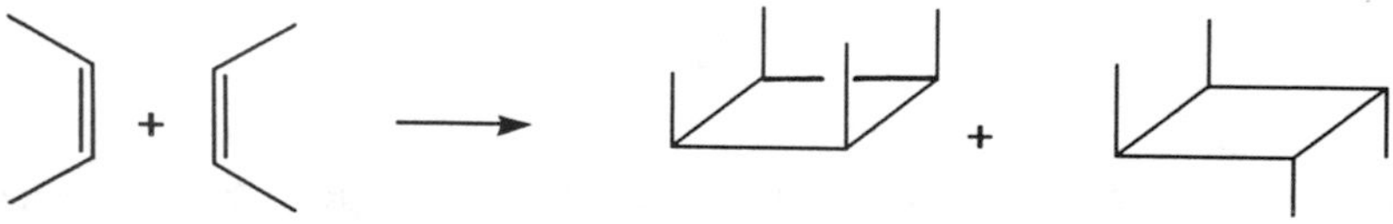

(6.3)

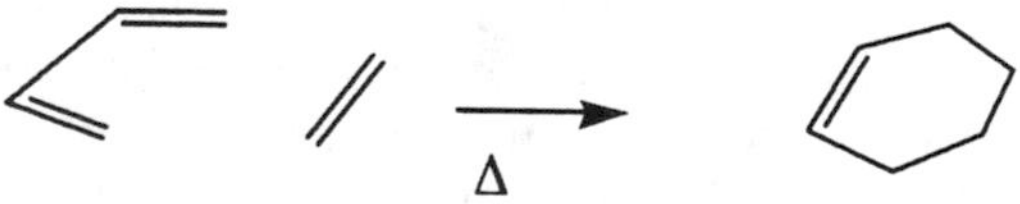

(6.4)

Sigmatropic reactions

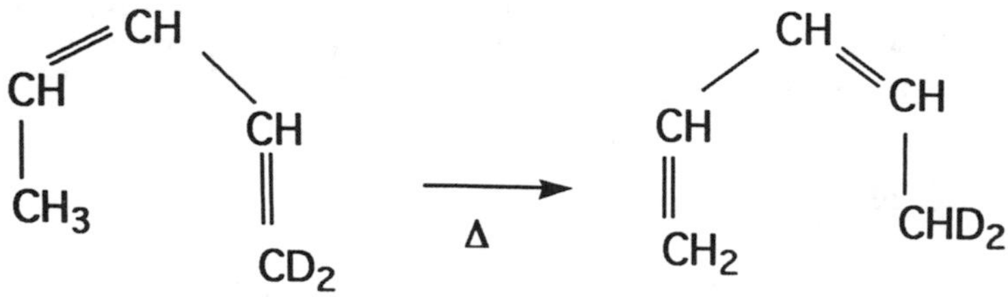

(6.5)

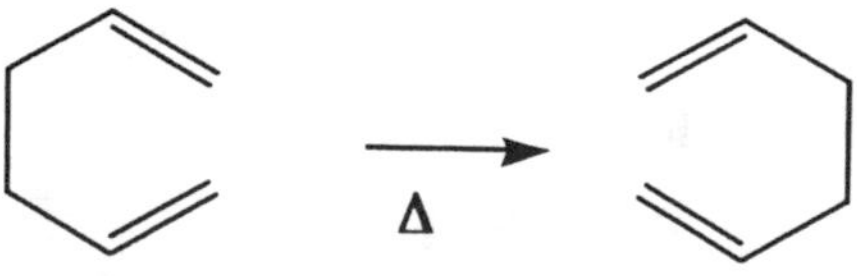

(6.6)

It will be obvious to organic chemists that the Diels-Alder reaction is a cycloaddition process, whereas the Cope and Claisen rearrangements are sigmatropic reactions.

Failure to produce concrete mechanistic evidence over the years proved frustrating to the point where it was once "a la mode" to refer to reactions of this type as "no mechanism" reactions. In 1965, however, all of this changed with the announcement of a series of rules by Professors Woodward and Hoffmann designed to predict the ease and steric course of all such concerted reactions. Indeed, the whole category was lumped into a more general type called "pericyclic reactions" proceeding through cyclic transition states formed by the continuous making of new bonds and breaking of the old.

The powerful predictive utility of the Woodward-Hoffmann rules has been thoroughly verified in the ensuing years, and even now the chemical literature is filled with a continuing spate of new and imaginative research suggested by the rules.

Woodward and Hoffmann originally grounded their rules in the so-called frontier orbital method and an extended form of HMO calculations. Subsequently, Longuet-Higgins and Abrahamson produced an argument supporting the rules based on the symmetry properties of the ground and transition state wave functions. Woodward and Hoffmann also utilized an orbital symmetry justification, which will be described in more detail later.

6.1 Evans's Rule

During the period marking the onset of World War II, M. G. Evans in England published several papers dealing with the mechanism of the Diels-Alder reaction. In these it was pointed out that the reaction was probably concerted and that a cyclic transition would possess the same stabilizing influence found in aromatic systems. Due to the war, Evans's work was largely ignored until it was returned to light by Professor Dewar.[1]

The concept of an aromatic transition state is particularly easy to deal with using PMO theory, for as we have seen in Chapter 3, the energy differences between cyclic and linear systems are readily ascertained. Dewar has generalized Evans's results to all pericyclic reactions, but before stating the generalization let us consider the case of the reaction of butadiene with ethylene (Eq. 6.4). The formation of cyclohexene requires that butadiene and ethylene collide to form some sort of a transition state, which then collapses, forming the products. Two possibilities exist, which are diagrammed on following page. Several points may be made regarding the diagrams. For one thing, we will continue to approximate π MOs as a linear combination of p AOs. The lobes of each AO are opposite in sign. In choosing the orbitals, it is customary to choose the AOs so that they overlap with the same sign or phase, positive with positive, etc. This choice is arbitrary, but it can be shown that the resultant MOs are independent of the signs of the basis set. It may be well to reiterate, however, that in Eq. 6.7 and in subsequent diagrams, the signs shown are associated with the basis set AOs.

Examination of A and B allows a comparison with hexatriene and benzene, respectively. For hexatriene one has six AOs and six electrons arranged in an

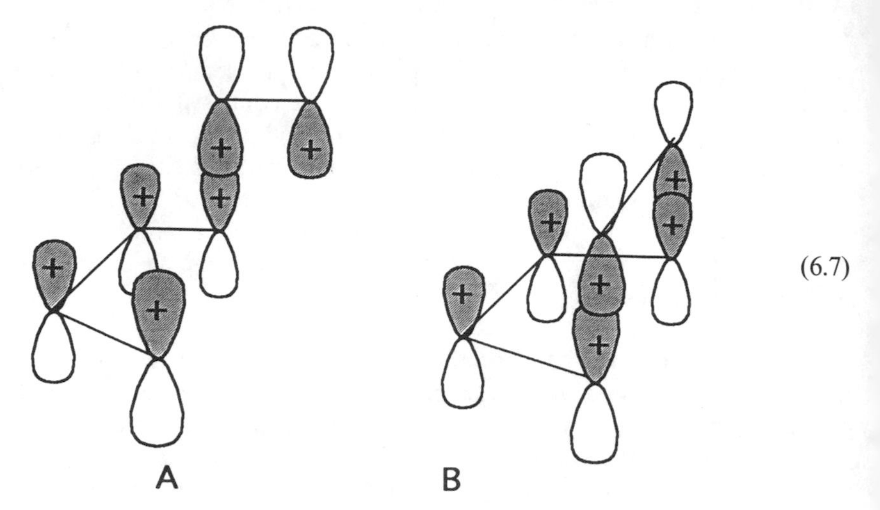

(6.7)

open-chain fashion, whereas in benzene the six AOs and π electrons comprise a cyclic system. Two systems are said to be isoconjugate if they have the same number of π electrons occupying π MOs of the same size and shape and formed by the interaction of the same number of AOs. Thus, A is isoconjugate with linear hexatriene, and B is isoconjugate with benzene. As shown previously, hexatriene is less stable than benzene, and rationally the preferred transition state for the Diels-Alder reaction will be the cyclic form B. The Diels-Alder reaction is an example of a (4 + 2) cycloaddition reaction.

The thermal dimerization of ethylene to form cyclobutane would be a (2 + 2) addition, which might similarly be expected to go through either a linear or a cyclic transition state. The cyclic transition state D is isoconjugate with cyclobutadiene and, therefore, is antiaromatic:

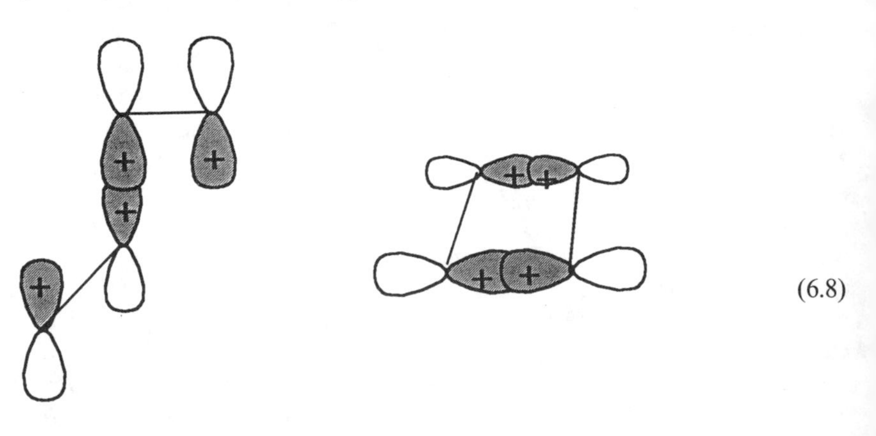

(6.8)

Such dimerizations do not occur readily, and where they do, a two-step diradical process is indicated. With the above examples behind us, it is now proper to give a more complete statement of Evans's rule: *Thermal pericyclic reactions take place preferentially through aromatic transition states.*

It should be added that the operting assumption in thermal reactions is that they occur with the reactant(s) in the ground state electronic configuration. More details must be supplied before Evans's rule can be extended to photochemical reactions.

6.2 Anti-Hückel Systems

Before proceeding to a more detailed consideration of pericyclic reactions, it is necessary to explore further the changes produced by altering the signs given the basis set AOs. As mentioned above, conventionally the AOs in ethylene are chosen to be in phase and the resonance integral is assigned the value β.

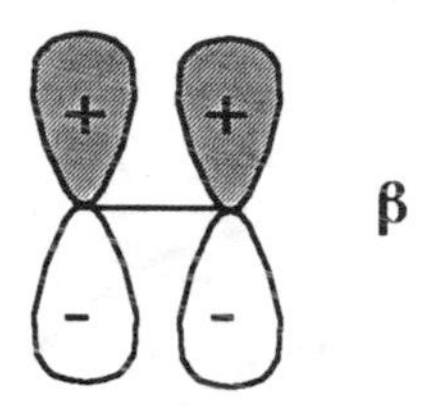

If one of these AOs is now twisted about the bond axis, the resonance integral will vary as the cosine of the twist angle. At 180° the value becomes $-\beta$.

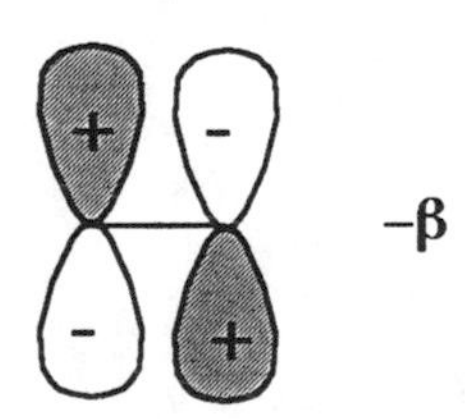

Intuitively one might expect this sign reversal to influence the π energies of polyenes calculated by the PMO method. However, this turns out not to be the case.

Consider for a moment the effect of phase dislocations introduced by twisting either an internal or a terminal AO in pentadienyl:

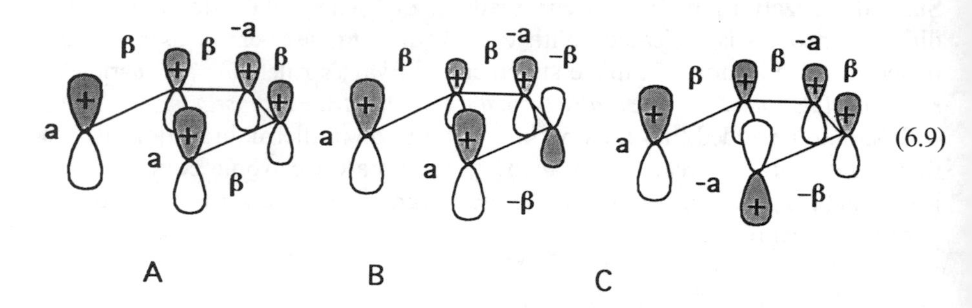

(6.9)

The first case (A) is the usual one. With the internal dislocation (B), two resonance integrals are reversed but the NBMO coefficients remain the same. Union of methyl with B will give the same results as with (A). Furthermore, in the case of Eq. 6.9C, union with methyl to form benzene would give $\delta E = 2[(a)\beta + (-a)(-\beta)] = 4a\beta$, which again is the value obtained from (A) and (B). Thus, as long as the number of changes in β around the ring is zero or an even number, the usual criteria for aromaticity apply.

Although no stable ring systems containing an odd number of changes in β can exist, it is interesting to consider how such a system might arise. Heilbronner has pointed out that an annulene with a half-twist in the ring would create a Möbius strip molecule with one phase dislocation:

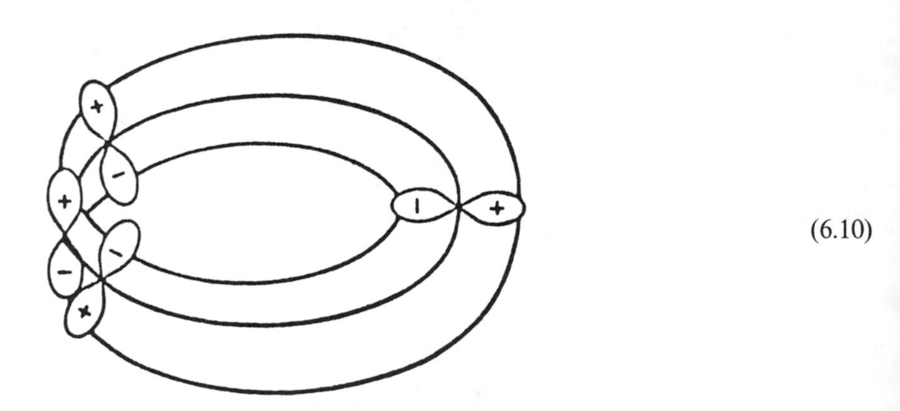

(6.10)

Considerations of the change in π energy on forming such a system reveal that now the $(4n + 2)$ systems are antiaromatic, whereas those with $4n$ π electrons are aromatic. Systems with an odd number of phase dislocations have been dubbed *anti-Hückel* systems. Their importance appears in transition states for certain pericyclic reactions.

A summary of what to expect from Hückel and anti-Hückel systems is given here, where (+) denotes aromatic and (−) antiaromatic character.

Table 6.1. The Relation of Ring Size and Structure to Aromatic Character.

	Rings					
	Even-Numbered			Odd-Numbered		
	$4n$	$4n+2$		$4n+1$ Cations	$4n+3$ Anions	
Hückel	−	+	−	+	+	−
Anti-Hückel	+	−	+	−	−	+

6.3 Electrocyclic Reactions

Electrocyclic reactions, as illustrated in Eq. 6.2, involve the interconversion of a conjugated polyene and the cyclic isomer formed by union between the terminal carbons. Since the reactions are usually reversible, we may apply the principle of microscopic reversibility, and for convenience of discussion consider a ring-opening reaction:

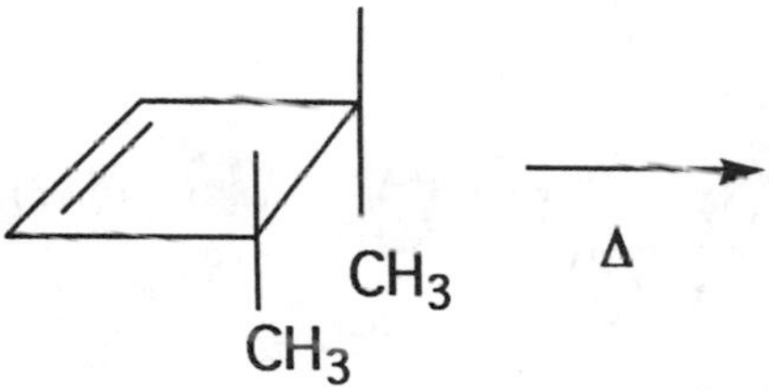

$$\text{(3,4-dimethylcyclobutene)} \xrightarrow{\Delta} CH_3CH=CHCH=CHCH_3 \quad (6.11)$$

Given that only one product is formed in a stereospecific fashion, it is evident that one of three geometries must result, i.e., *cis, cis* (A), *trans, trans* (B), and *cis, trans* (C):

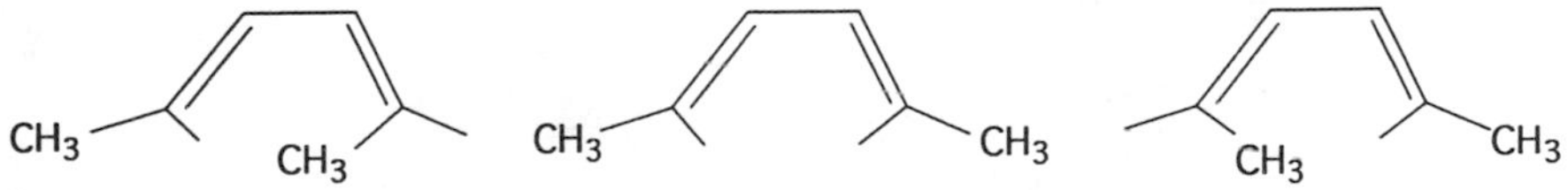

A B C

Clearly, the ring opening must be accompanied by rotation of the methyl groups into their appropriate planar configurations. Woodward and Hoffmann have termed the two possible modes *disrotatory* and *conrotatory*, depending on the sense of the rotation about the developing double bonds.

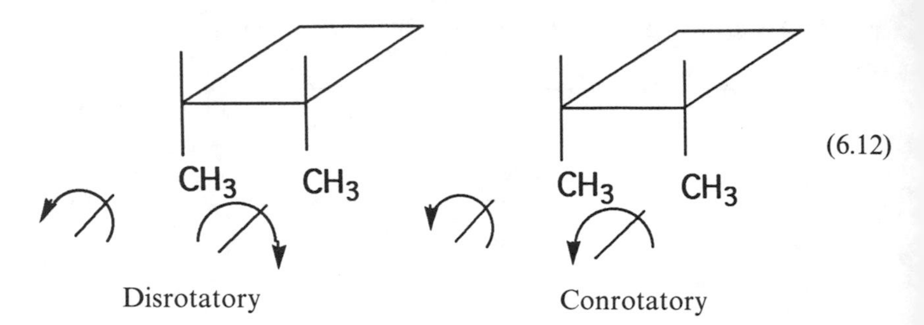

(6.12)

Examination shows that (A) and (B) can arrive only via disrotatory processes, whereas (C) results from a conrotatory process.

The transition states for the ring opening will require the two carbons bearing the methyls to develop sp^2 character as the sigma bond ruptures. The transition state will be isoconjugate with cyclobutadiene with four p orbitals and four π electrons. In the case of disrotatory opening, a normal Hückel system will be developing so that the transition state will be antiaromatic. However, for the conrotatory process an anti-Hückel system with one phase dislocation is formed, and this will be aromatic. Obviously the conrotatory process is favored, and one does isolate only the *trans*, *cis* product (C) from the reaction.

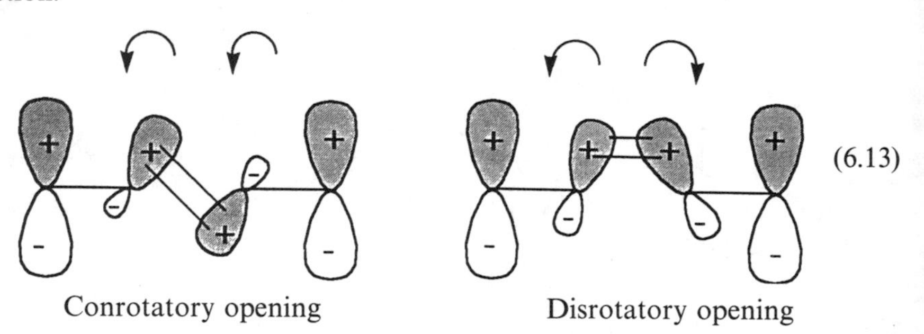

(6.13)

Consideration of the two possible transition states for the thermal ring closure in Eq. 6.2 shows the disrotatory process to lead to a transition state isoconjugate with benzene.

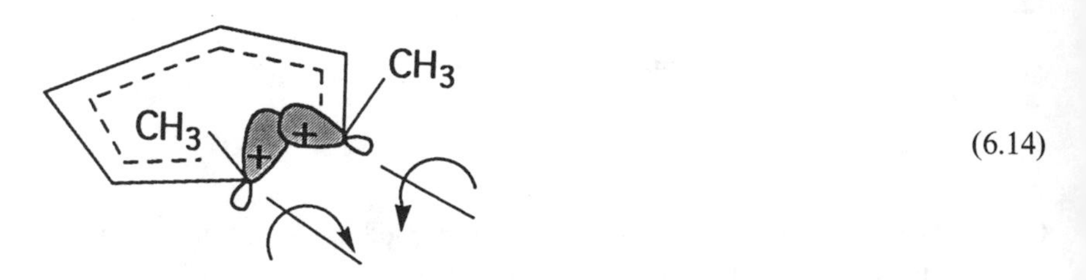

(6.14)

Thus, thermal cyclizations of 1,3,5-trienes will preferentially occur via the aromatic $(4n + 2)$ transition state resulting from the disrotatory process.

6.4 Photochemical Reactions

As demonstrated in Eq. 6.2, pericyclic reactions may take place through the action of light. Often the products are different from those produced thermally. Usually this difference appears in the stereochemistry of the product.

It is now well known that photochemical reactions occur through excited reactants; singlets and triplets formed by promoting an electron from the highest occupied orbital (HOMO) of the ground state to the lowest unoccupied orbital (LUMO).

The PMO method is not capable of distinguishing singlet and triplet states. However, this fact appears to bear no consequences for qualitative arguments on photochemical pericyclic reactions.

Although we will not do so here, it is possible to make use of arguments similar to those presented previously regarding the stabilities of transition states to show that photochemical reactions proceed in just the opposite sense of thermal reactions, i.e., *Photochemical pericyclic reactions take place preferentially through excited forms of antiaromatic transition states.*

This means that the conclusions regarding the permissiveness and stereochemistry of photochemical reactions will be just the reverse of those for the corresponding thermal reactions.

Thus, the (2 + 2) cycloaddition reactions of olefins to form cyclobutanes, although thermally disallowed, are photochemically allowed. Indeed, the photochemical pathway is often the synthetic method of choice in the preparation of substituted cyclobutanes (see Eq. 6.3).

For the photochemical electrocyclic reaction in Eq. 6.2, the $(4n + 2)$ atoms of the transition state must show a phase dislocation, since in order to be antiaromatic it must also be anti-Hückel (see Table 6.1). Hence, the transition state will look like

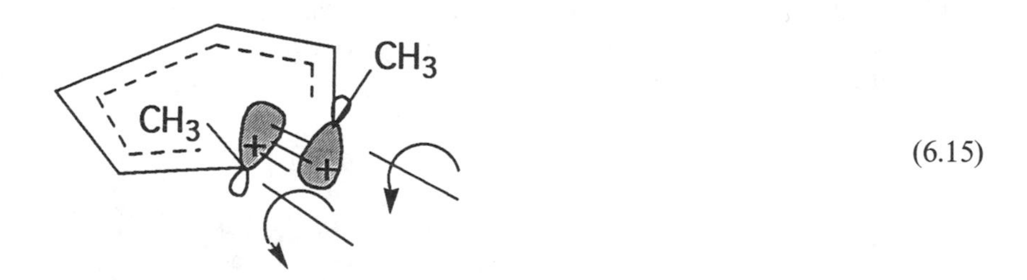

(6.15)

6.5 Sigmatropic Rearrangements

These reactions are defined by the migration of a σ bond flanked by one or more π bonds to a new position in the molecule.

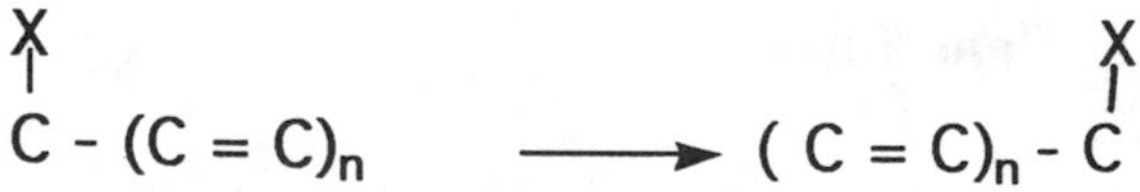

$$\overset{X}{\underset{|}{\uparrow}}\!C - (C = C)_n \longrightarrow (C = C)_n - \overset{X}{\underset{|}{\uparrow}}\!C \tag{6.16}$$

In addition to the examples 6.5 and 6.6 given previously, two other examples will be given here:

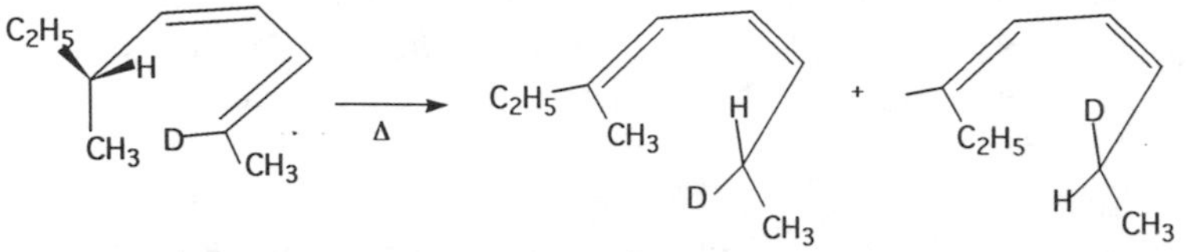

(6.17)

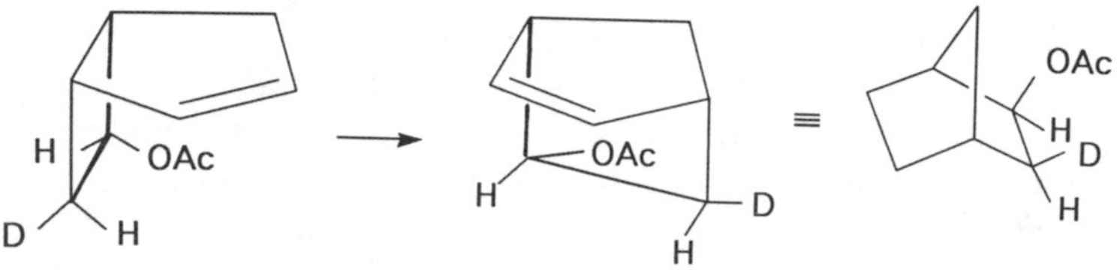

(6.18)

It will be more convenient if two matters of nomenclature are cleared up before proceeding. First, the reaction is termed an $[i, j]$ sigmatropic rearrangement when the bond migrates from position [1, 1] to position $[i, j]$. Reaction 6.5 is a [1, 5] rearrangement, whereas reactions 6.6, 6.17, and 6.18 are [3, 3], [1, 5], and [1, 3] rearrangements, respectively.

Secondly, if the migrating group remains on the same side of the π system, the process is said to be *suprafacial*. Migrations that cross the nodal plane of the π system are called *antarafacial*.

Consider the transition states for the [1, 3] rearrangement of hydrogen. The structure (A) is isoconjugate with cyclobutadiene and, as a Hückel system, is antiaromatic and unstable. Although the antarafacial, anti-Hückel transition state (B) is energetically favorable, the span for H orbital overlap is prohibitively large:

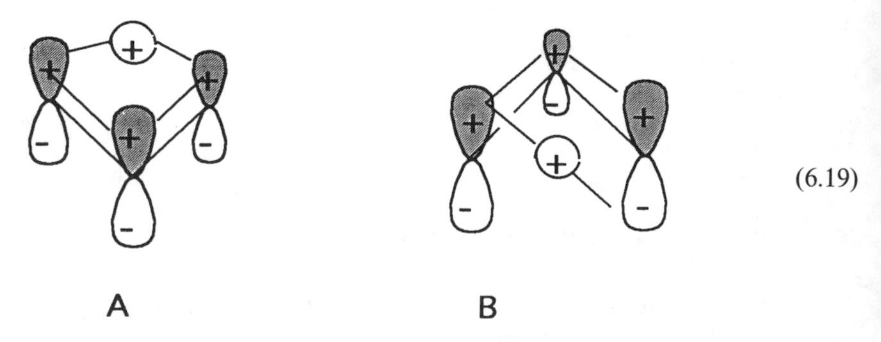

(6.19)

Thus, thermal [1, 3] rearrangements of hydrogen are not observed.

However, in Eq. 6.18 the rearrangement may proceed through the anti-Hückel system in Eq. 6.20, which is not only aromatic but sterically possible,

(6.20)

The [1, 5] rearrangements of Eqs. 6.5 and 6.18 proceed through suprafacial Hückel ($4n + 2$) transition states.

6.6 The Conservation of Orbital Symmetry[2]

The basis of the approach of Woodward and Hoffmann is the concept that the MOs of the product must be generated smoothly from MOs in the reactants having the same symmetry properties. Thus, for a thermal pericyclic reaction, the electrons of the reactant(s) will be in the ground state MOs. If these MOs are converted into MOs of the product without a change in symmetry, the reaction is allowed, i.e., energetically favorable. For photochemical reactions, the reactant is in the first excited state. Again the differentiation between singlet and triplet states is ignored.

Several approaches to the Woodward-Hoffmann rule have been made. The most popular and utilitarian appears to be the use of correlation diagrams, and as a comparison to the PMO approach we will briefly examine how these are drawn. Briefly, the task is to write down the approximately known energy levels of the reactants on one side and the products on the other. Assuming a geometry of approach, one classifies both sets according to the symmetry maintained during the approach, Levels of like symmetry are then connected, following the rule that only levels of unlike symmetry are allowed to cross.

Consider again the union of two ethylenes to form cyclobutane. In all, four MOs are involved (a bonding and an antibonding for each ethylene) for the reagents, and similarly there will be four new σ orbitals in the product. The Woodward-Hoffmann rule requires that for the reaction to occur in a concerted fashion, the orbitals of the reactants must pass into those of the products in a continuous manner, i.e., the symmetry of the orbitals must be conserved. The approach of two ethylenes to form the cyclic transition state is characterized by two symmetry planes:

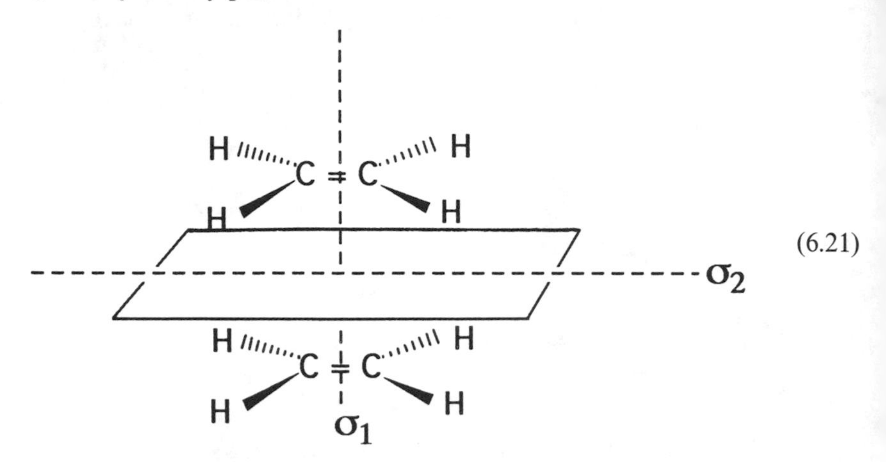

(6.21)

The interaction of the two sets of ethylene MOs can be expressed in terms of these symmetry planes and the phases of component atomic orbitals. The designations indicate whether the orbitals are symmetric (S) or antisymmetric (A) about the symmetry planes σ_1 and σ_2, respectively. If the wave functions are taken as ψ_1 and ψ_2 for the ground states and ψ_1^* and ψ_2^* for the antibonding states, then the wave functions for the reactants in the diagrams below can be written as shown:

SS SA AS AA

σ_2

σ_1

(6.22)

$$\Psi_1 = \frac{1}{\sqrt{2}}(\psi_1 + \psi_2) \quad \text{SS}$$

$$\Psi_2 = \frac{1}{\sqrt{2}}(\psi_1 - \psi_2) \quad \text{SA}$$

$$\Psi_3 = \frac{1}{\sqrt{2}}(\psi_1^* + \psi_2^*) \quad \text{AS}$$

$$\Psi_4 = \frac{1}{\sqrt{2}}(\psi_1^* - \psi_2^*) \quad \text{AA}$$

Since the π systems are independent in the reactants, E_1 and E_2 are equal as are E_3 and E_4.

In progressing to the new σ bonds of the product, the symmetry must be preserved under the rule. New wave functions for the two σ bonds may be defined as below:

SS SA AS AA

σ_2 (6.23)

σ_1

$$\Phi_1 = \frac{1}{\sqrt{2}}(\phi_1 + \phi_2) \quad \text{SS}$$

$$\Phi_2 = \frac{1}{\sqrt{2}}(\phi_1 - \phi_2) \quad \text{SA}$$

$$\Phi_3 = \frac{1}{\sqrt{2}}(\phi_1^* + \phi_2^*) \quad \text{AS}$$

$$\Phi_4 = \frac{1}{\sqrt{2}}(\phi_1^* - \phi_2^*) \quad \text{AA}$$

where φ represents the newly formed σ bond wave functions formed from carbon sp^3 atomic orbitals. Since the two σ bonds are independent in the product, again E_1 will equal E_2 and E_3 will equal E_4. We can now plot the relative energies of the various orbitals as follows:

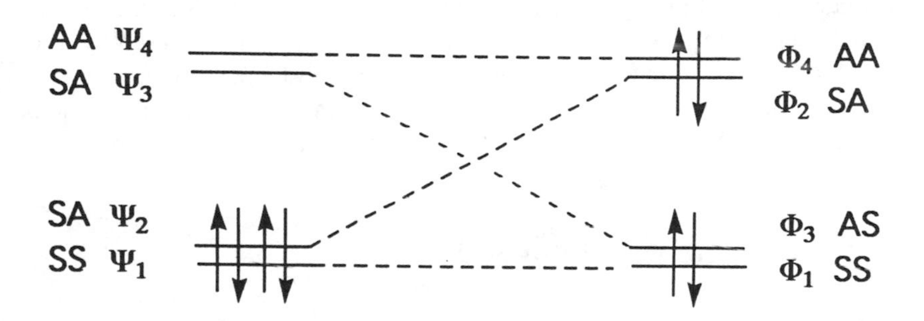

Note that it is φ_3, an AS function, which becomes a bonding MO in the product, while the MO associated with φ_2 is antibonding. It follows under the Woodward-Hoffmann rule that for a concerted (2 + 2) cycloaddition where orbital symmetry is preserved, a high-energy transition state leading to an electron pair in an antibonding orbital of the product is formed. Such a process is considered energetically forbidden. If, however, one of the reagent electrons is promoted to ψ_3, then the formation of a diradical product becomes thermochemically allowed. Such a promotion can be brought about by the absorption of a photon.

An entirely analogous process may be carried out for a (4 + 2) cycloaddition reaction, such as the butadiene plus ethylene reaction. The same symmetry elements exist for this system as for that above. The correlation diagram is

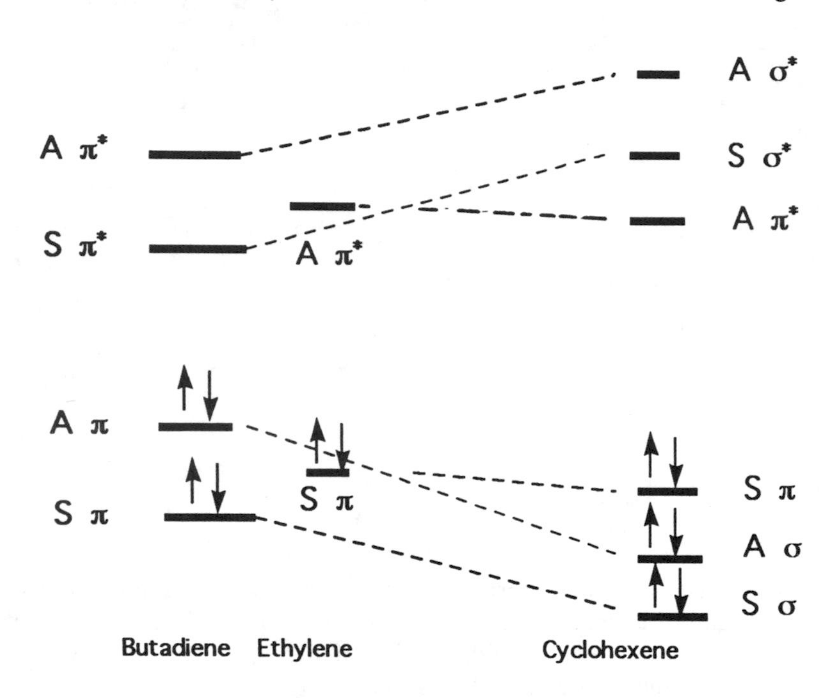

Here symmetric and antisymmetric π orbitals are designated Sπ and Aπ, with the stars reserved for the antibonding MOs. Similar notation is used for the σ bonds in the product. The MOs for ethylene and butadiene are given on pages xx and yy. Since the preservation of orbital symmetry leads only to a net lowering of energy, this process is thermochemically allowed.

6.7 Electrocyclic Reactions—2

In order to apply orbital symmetry considerations to electrocyclic reactions, it is necessary to consider the consequences of both conrotatory and disrotatory processes. Again, let us consider the interconversions of cyclobutenes and butadienes. It will be necessary to consider the rupture of the cyclobutene σ bond and the π system as they are converted into butadiene orbitals. Let us consider these aspects independently with reference to the symmetry axis of the molecule and for the conrotatory process:

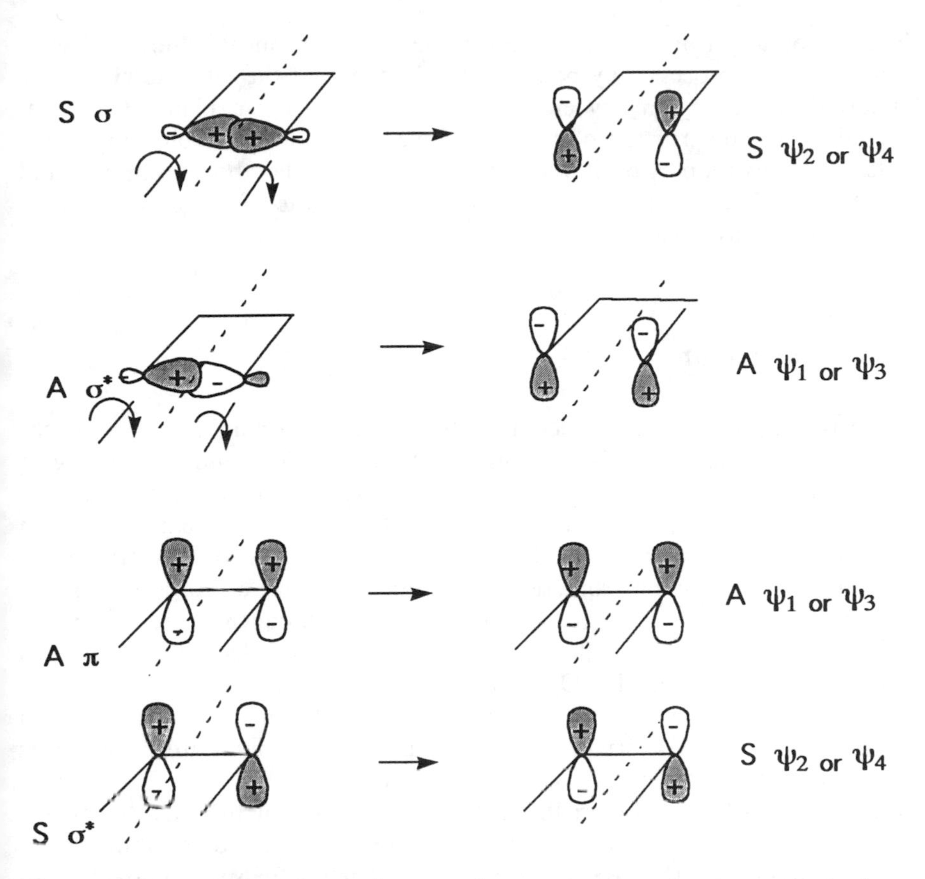

The symmetry designation refers to the symmetry axis shown and those butadiene orbitals which correspond to the symmetry of the AOs shown. An energy correlation diagram may now be drawn:

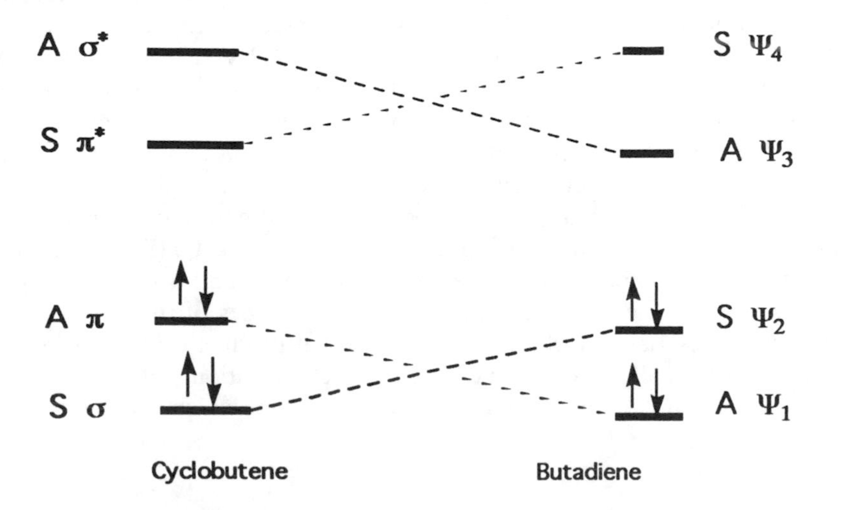

The conrotatory ring opening is seen to be thermochemically allowed. Essentially the same process may be followed to show that the disrotatory process is forbidden, the symmetry element in this case being a plane. The point regarding the construction of correlation diagrams and the application of the rule of conservation of orbital symmetry is made in the above examples, and more detailed texts on the subject should be consulted for the extension to sigmatropic rearrangements.

6.8 FMO Theory[3]

Concepts derived from molecular orbital theory pertaining to pericyclic reactions developed rapidly in the 1960s and 1970s. Short of plunging back into the original literature, it is difficult to reconstruct the exact sequence in which these concepts were introduced. Certainly one of the early players in this fascinating game was the K. Fukui who applied PMO theory to the question of electrophilic aromatic substitution and went on to develop much of frontier molecular orbital (FMO) theory.[3a] Subsequently, Klopman expanded the theory and popularized it in his 1974 book.[3c] Dewar has criticized certain applications of FMO theory as providing inaccurate predictions, preferring PMO theory.[1a] However, one still finds much terminology and application of FMO theory in today's organic chemistry literature, so chemists obviously still find the concepts of FMO theory of use. Although FMO theory applies to many kinds of organic reactivity, the present brief treatment will confine itself to the subject of pericyclic reactions. Readers who wish a more complete treatment are referred to the delightful small book by Ian Flemming.[3b]

The basic premise of FMO theory is that reactions are controlled by the energetics of the highest occupied (HOMO) and the lowest unoccupied (LUMO) molecular orbitals. The energetic consequences of two orbitals coming together may be summarized as in Figure 6.2. Again, the interacting orbitals must have the same phase relation, as two unlike phase overlaps lead to an increase in energy. At the top of Figure 6.2, the interaction of two filled HOMO levels is shown as having a very small net effect. The gain of one filled product orbital is offset by the increased energy of the higher orbital. The analogy would be that of attempting to react two helium atoms with each other. When the HOMO and LUMO (Fig. 6.2, center) are comparable in energy, the interaction is strongest, and a net lowering of energy occurs. As the separation of the relative energies of the HOMO and LUMO increases, the perturbation to the resultant product orbitals decreases, with a resultant small stabilization due to product formation. Let us now apply these concepts to two cycloaddition reactions with which we have some familiarity.

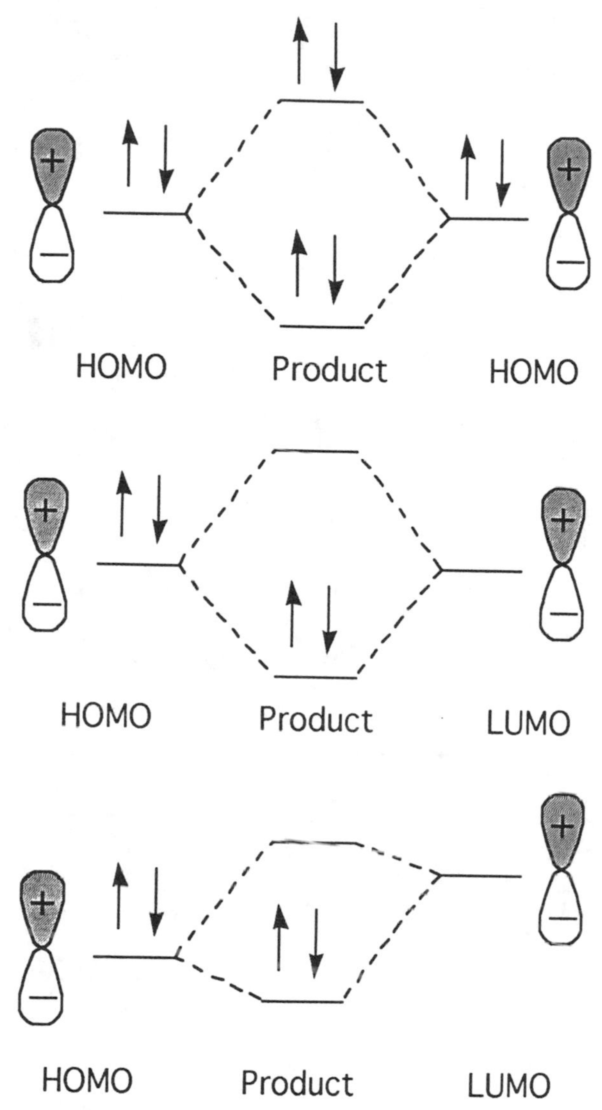

Figure 6.1. Effects of bringing two reagents together.

The diagram on following page shows the HOMO-LUMO system for butadiene and ethylene. At the top the HOMO for the diene interacts with the LUMO for the dienophile, whereas on the bottom the situation is reversed. The HOMO-LUMO gap is the same in either case. There is a net energy lowering and the phases are correct in either case. The reaction is allowed and is suprafacial as shown.

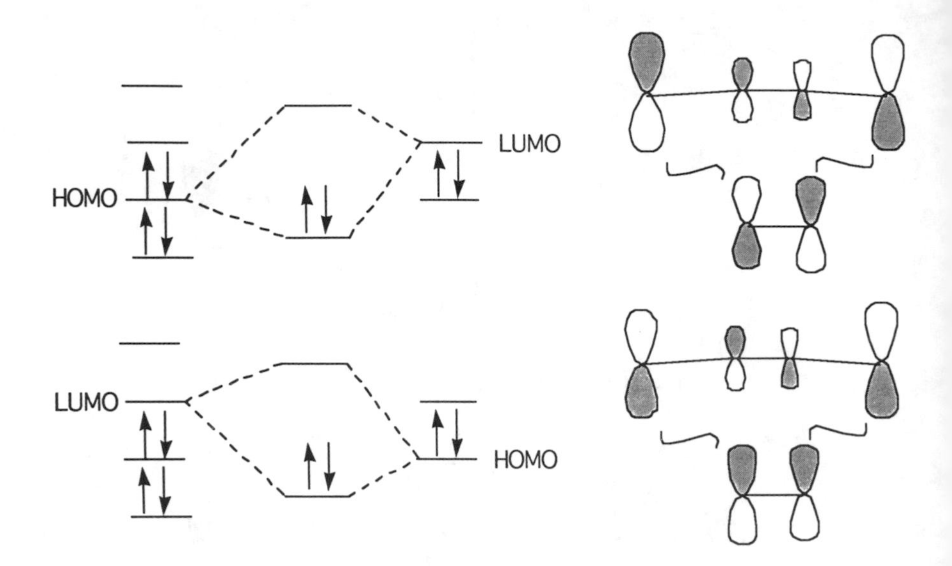

On the other hand, the situation for the thermal (2 + 2) cycloaddition is shown below. Although it appears the energetics are still favorable for this case, it can be seen that the ring is forming in an antarfacial fashion:

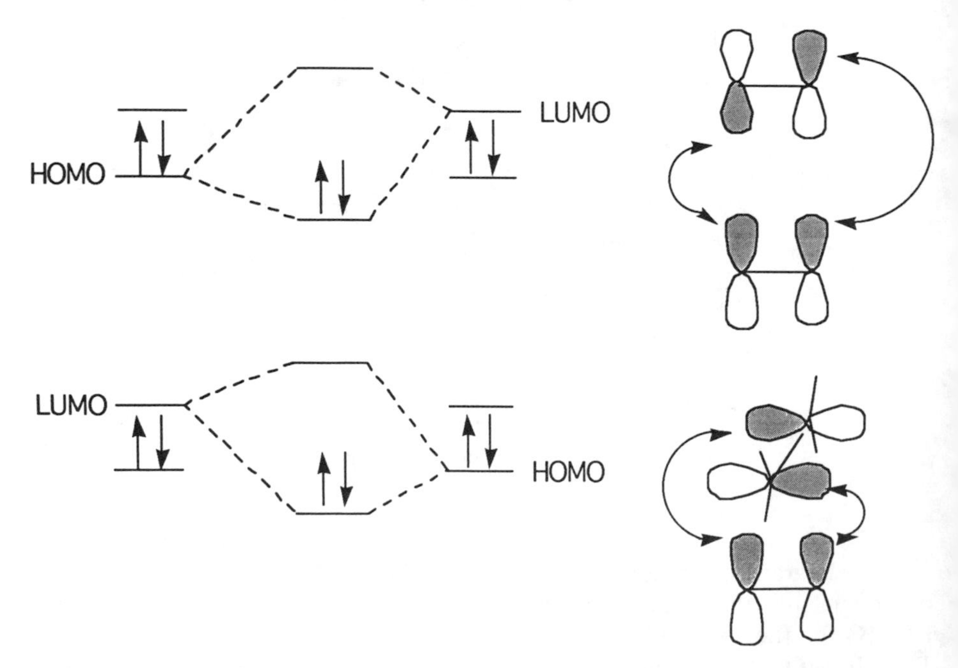

This is equivalent to forming a twisted transition state (lower right)—a form that has never been observed in a (2 + 2) cycloaddition reaction. In the twisted form, the substituents on the top ethylene would clearly prevent the two π

systems from properly overlapping. The (4 + 2) reaction is *symmetry-allowed*, whereas the (2 + 2) reaction is *symmetry-forbidden.*

The situation changes for the (2 + 2) photochemical process, in that now one ethylene has been placed in its first excited state. This creates

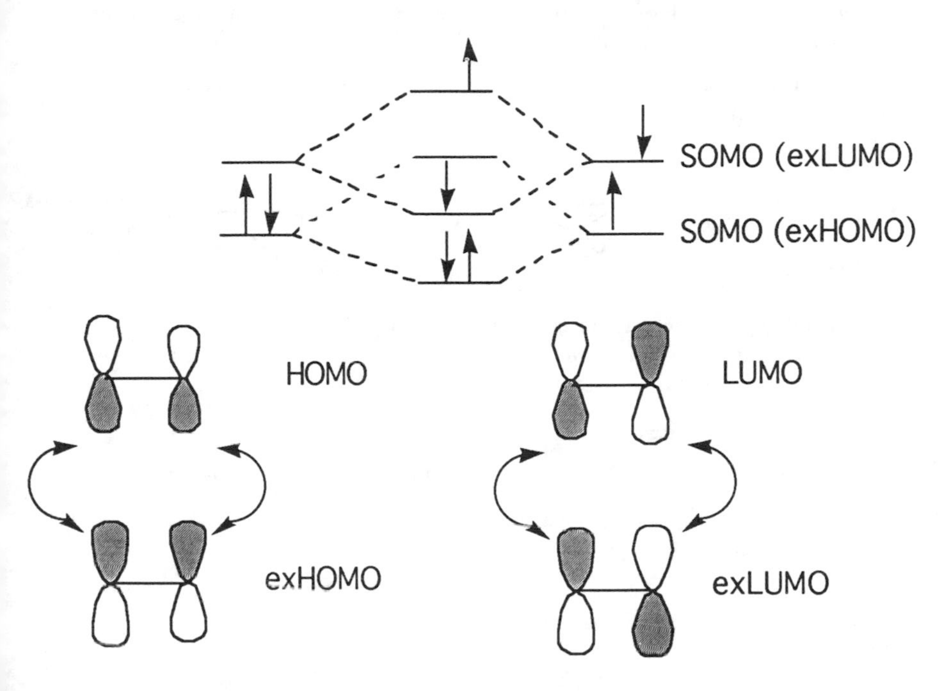

two singly occupied molecular orbitals (SOMOs), which are required to interact with their counterparts in the ground state olefin, i.e., the HOMO interacts with the exHOMO and the LUMO with the exLUMO. As shown above, these interactions are *symmetry-allowed.* Of course, for the Diels-Alder reaction it is possible to show that the photochemical process is *symmetry-forbidden.* Examples applied to the other types of pericyclic reactions can be found in the text by Flemming.[3b]

6.9 Conclusion

Which approach is better in treating pericyclic reactions? In most cases the three approaches are in agreement, and the practitioner is left to choose that scheme with which he is most comfortable. The author has felt that the PMO approach was generally easier to learn and apply, but others may disagree. Dewar has produced an argument that the PMO theory is on a more sound theoretical basis than the Woodward-Hoffmann and FMO approaches.

Problems

1. What stereochemistry would you expect from the thermal ring opening reactions of *cis* and *trans* 5,6-dimethyl-1,3-cyclohexadiene?

2. Demonstrate for each of the products in Eq. 6.17 whether the hydrogen migrates in a suprafacial or an antarafacial manner.

3. Discuss the mode of migration in cases and show them to be consistent with the PMO approach.

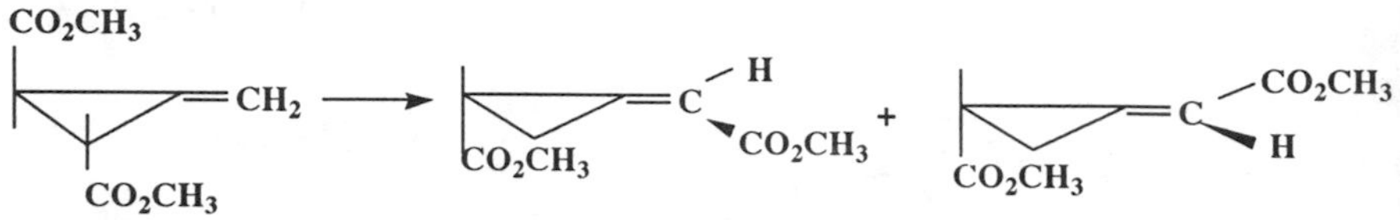

4. Upon heating, the deuterated compound on the left suffers scrambling. Since one cannot separate the isomers, the effect is as though there were multiple labeling, as indicated in the equation. Account for this observation.

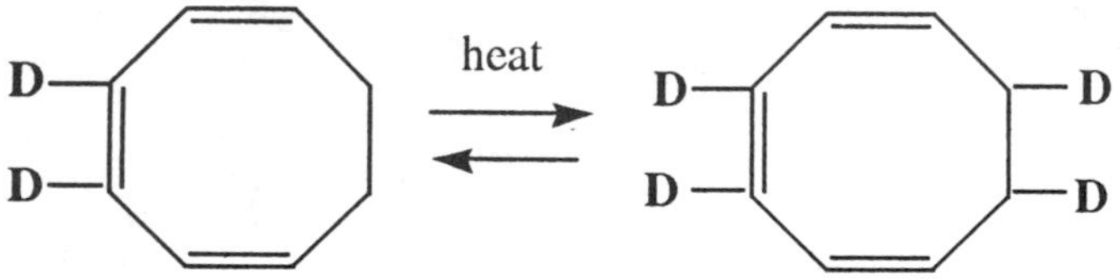

References

1. (a) M. J. S. Dewar, *The Molecular Orbital Theory of Organic Chemistry*, McGraw-Hill, New York, 1969; (b) M. J. S. Dewar and R. C. Dougherty, *The PMO Theory of Organic Chemistry*, Plenum Press, New York, 1975.

2. (a) R. B. Woodward and R. Hoffmann, *The Conservation of Orbital Symmetry*, Academic Press, New York, 1970; (b) A. P. Marchand and R. E. Lehr, *Pericyclic Reactions*, Vols. I and II, Academic Press, New York, 1977.

3. (a) K. Fukui, *Acc. Chem. Res.*, **4**, 57 (1971); (b) I. Flemming, *Frontier Orbitals and Organic Chemical Reactions*, 2nd ed., Cambridge University Press, 1979; G. Klopman (ed.), *Chemical Reactivity and Reaction Paths*, John Wiley & Sons, New York, 1974.

CHAPTER

7

Improvements and Extensions of the Hückel Theory

7.1 An Introduction to Self-Consistent Methods

In many ways it is remarkable that the Hückel theory provided the wide range of useful results that we have just seen demonstrated. The theory deals only with π electrons in planar conjugated molecules. HMO theory is based on a one-electron treatment, i.e., the orbital equations are developed for a one-electron Hamiltonian, and then two electrons having opposite spins are placed in each. No interactions between the electrons are taken specifically into account, and no integrals are ever evaluated. As a consequence, HMO theory cannot differentiate the energies of singlet and triplet states. In the most naive early versions of HMO theory, Coulomb and bond integrals were each given single values throughout, resulting in difficulties with charged systems such as ions and non-AHC systems. This will also produce faulty results for systems where bond lengths vary.

A conspicuous failure of HMO theory mentioned earlier is the exaggeration of charge densities in non-AHC. When $q \neq 1$, it is illogical to assume that $\alpha = \alpha_0$. A net positive charge would mean a reduced nuclear screening charge for electrons and an increased Coulombic attraction. The opposite is true if the net charge is negative. This will be true also when other charged entities are present in the molecule. Wheland and Mann first addressed this problem in 1949.[1] They suggested a change that may be written as

$$\alpha_{n+1} = \alpha_n + (1 - q_n)\omega\beta_0$$

For obvious reasons, this method has been called the ω *technique*. A range of values for the adjustable parameter ω has been suggested, but Streitwieser[2] has suggested the value 1.4 to be generally the most applicable. To apply the ω technique, one first conducts a standard HMO computation using these results to calculate the charge densities. These original q values are then used in the equation above to generate a new set of α values. With these in hand, new values of q_i are generated, and the iteration continues until convergence is reached. Such calculations are said to be *self-consistent*. Streitwieser has calculated the charge distribution for the benzyl cation by the ω technique. His values are here compared with our previous values from the simple HMO theory. As can be seen, the effect is to even out the extremes of charge predicted by HMO theory.

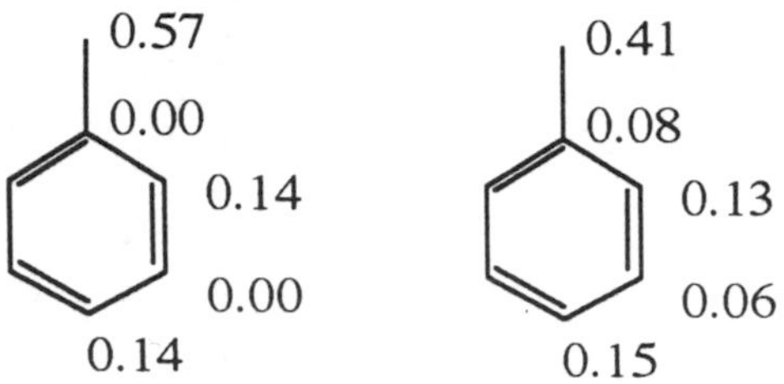

0.57 0.00 0.14 0.00 0.14

0.41 0.08 0.13 0.06 0.15

Charge densities HMO ω technique

Methods that lead to self-consistent results will play a very important role in many of the calculations to be discussed.

It is reasonable to expect the bond integral β to vary with bond length (and bond order). Brogli and Heilbronner[3] established a relation similar to that for α above relating the value of β to bond length. The iterative process was again used to achieve self-consistency. They found that orbital energies, when adjusted for changes in β, correlated very well with ionization potentials for a large series of AHCs and non-AHCs.

One of the major curiosities of HMO theory is the fact that electron density builds up in bonding regions of the molecule, yet the simplifying assumption is made that there is no overlap between the AOs that make up the MO. We have seen in Eq. 1.14 that the energy term for the energy levels contains the expression $1 + S_{12}$ in the denominator. Remembering that both α and β are negative quantities, the antibonding orbital is higher above the $E = \alpha$ line than the bonding MO is below the line. The inclusion of overlap destroys the symmetry of the orbitals with reference to this line. As a consequence, if it were possible to fill both of the orbitals in the hydrogen molecule, there would be a net antibonding energy balance, and the molecule would be incapable of existence. When a value for S_{ij} is expressly included in the calculations for AHC π systems, the symmetry of the orbitals about the $E = \alpha$ line is similarly destroyed. It can be shown that even though the value of E_{π}^{total} is altered, the effect is just a scale factor. The failure to specifically include overlap in HMO calculations is justified because qualitative results are not materially affected.

McWeeny[4] has provided an interesting derivation bearing on the approximation that results from neglecting the overlap integral. In part, this is based on earlier work by Löwdin.[5] Since this touches on the later semiempirical methods discussed in Chapter 9, it is instructive to summarize his approach. We will follow the format used in Chapters 1 and 2.

The secular equations for a system of two AOs (ψ_1 and ψ_2) may be written as

$$(\alpha - E)C_1 + (\beta - ES)C_2 = 0$$

$$(\beta - ES)C_1 + (\alpha - E)C_2 = 0$$

Now we define two new AOs in which ψ_1 has a small amount of ψ_2 subtracted, and its counterpart for ψ_2, i.e.,

$$\bar{\psi}_1 = \psi_1 - \lambda\psi_2 \qquad \text{and} \qquad \bar{\psi}_2 = \psi_2 - \lambda\psi_1 \tag{7.1}$$

the MO formed by the LCAO can then be written as

$$\Psi = C_1\psi_1 + C_2\psi_2; \text{ or equally well as } \bar{C}_1\bar{\psi}_1 + \bar{C}_2\bar{\psi}_2 \tag{7.2}$$

If we now take (and equating the coefficients of ψ_1 and ψ_2)

$$\bar{C}_1 - \lambda\bar{C}_2 = C_1 \qquad \text{and} \qquad -\lambda\bar{C}_1 + \bar{C}_2 = C_2$$

which may be solved to give

$$\bar{C}_1 = \frac{C_1 + \lambda C_2}{1 - \lambda^2} \qquad \text{and} \qquad \bar{C}_2 = \frac{C_2 + \lambda C_2}{1 - \lambda^2} \tag{7.3}$$

This conversion of one basis set to a related basis where the new coefficients can be expressed in terms of the old is called a *transformation.* The transformation here provides that, given a correct value for λ, the new basis set will be orthogonal. This condition is met by the following relation:

$$\int (\psi_1 - \lambda\psi_2)(\psi_2 - \lambda\psi_1)\, d\tau = S - \lambda - \lambda + \lambda^2 S = 0$$

As both S and λ are small, the last term may be ignored, with the result that $\lambda \cong \frac{1}{2}S$, and neglecting terms in S^2, the new orthogonalized wave functions are

$$\bar{\psi}_1 = \psi_1 - \tfrac{1}{2}S\psi_2 \qquad \text{and} \qquad \bar{\psi}_2 = \psi_2 - \tfrac{1}{2}S\psi_1$$

which leads to the new secular equations

$$
\begin{aligned}
(\bar{\alpha} - E)\bar{C}_1 + (\bar{\beta})\bar{C}_2 &= 0 \\
(\bar{\beta})\bar{C}_1 + (\bar{\alpha} - E)\bar{C}_2 &= 0
\end{aligned}
\tag{7.4}
$$

where the new Coulomb and bond integrals are defined as

$$\bar{\alpha} = \int (\psi_1 - \lambda\psi_2)H(\psi_1 - \lambda\psi_2)\, d\tau = (1 + \lambda^2)\alpha - 2\lambda\beta$$

and

$$\bar{\beta} = \int (\psi_1 - \lambda\psi_2)H(\psi_2 - \lambda\psi_1)\, d\tau = (1 + \lambda^2)\beta - 2\lambda\alpha$$

For $\lambda \cong \frac{1}{2}S$ (ignoring terms in λ^2),

$$\bar{\alpha} \simeq \alpha - S\beta \qquad \text{and} \qquad \bar{\beta} \simeq \beta - S\alpha \tag{7.5}$$

The implications of the relations in Eq. 7.5 are that we may set up a series of secular equations such as Eq. 7.4 and solve for the MOs and energies without the need for including overlap expressly. The results can then be translated by utilizing the new definitions in Eq. 7.5 for parameters α and β. The complete or partial neglect of overlap becomes an important consideration in what is to follow.

7.2 Extended Hückel Theory (EHT)

Following World War II the first computers began to appear in government and university laboratories. Immediately theoreticians started on the problem of extending quantum mechanical calculations to many-electron problems involving both σ and π electron systems and no longer confined to planar conjugated molecules.

Not the first, but perhaps the most successful, of these early attempts was made by Hoffmann at Cornell.[6] Equations 2.5 and 2.6 describe the general format for the secular equations and secular determinant; they are repeated here for convenience:

$$\sum_{i=1}^{n} c_i[H_{ij} - ES_{ij}] = 0 \qquad j = 1, 2, 3, \ldots, n \tag{2.5}$$

$$|H_{ij} - S_{ij}E| = 0 \tag{2.6}$$

Hoffmann followed much the same line of argument as Hückel but applied the above equations to saturated as well as all-electron π systems. "All-electron" means that electrons in σ as well as π bonds are included. Only the valence electrons for the various atoms are to be considered. Inner shell electrons were averaged in with the nuclear charge in the "core approximation," the magnitude of which was assigned by parameterization. In the case of methane, a linear combination of four hydrogen $1s$ orbitals with carbon $2s$, $2p_x$, $2p_y$, and $2p_z$ orbitals is employed.

So far we have neatly sidestepped any account of how to attribute a mathematical form to the description of orbitals for elements beyond hydrogen in the periodic table. It is quite possible to write a Hamiltonian for a nucleus with several electrons around it, but it is quite impossible to solve exactly the resulting wave equation, because the charge distribution of any one electron depends on the charge distribution of all of its neighbors. A solution suggested by Hartree in 1928 has been widely adapted to many-electron problems. The solution can be described as follows: Wave functions for all the electrons but one are assumed, and the electron density distribution is then calculated. The wave equation for the one electron is now written including the electron distribution as approximated in the first step. The wave function for the one electron is then calculated. The procedure is then repeated for a second and all other electrons. The resultant new wave functions are used in a second cycle to derive an improved set of wave functions and energies. The calculation is repeated until a stable energy minimum is found. Such functions are called self-consistent field (SCF) wave functions.

Following this logic, Slater produced an approximate set of atomic wave functions (Slater-type orbitals, STO hereafter), which may be written as follows for atoms in the first two rows of the periodic table. The wave functions for the $2p_y$ and $2p_z$ orbitals follow by analogy:

$$\psi(1s) = N_{1s}e^{-cr}, \quad \psi(2s) = N_{2s}e^{-cr/2}$$

$$\psi(2p_x) = N_{2p}xe^{-cr/2}$$

The bond radii r are expressed in the bohrs units (0.529 Å), and the effective nuclear charge c is determined by a set of rules that are not pertinent to our further interests. The N values are a set of empirical parameters.

For the initial extended Hückel calculations, STOs were chosen with the appropriate prefix coefficients. No hybridized orbitals were used in the computation. For methane, there will be four AOs for the carbon and one each for the four hydrogens. The secular equations will contain eight terms, each leading to an 8×8 secular determinant of the form overleaf. This determinant is still a one-electron LCAO treatment, as in the original Hückel approximation. The hydrogen $1s$ orbitals are given by subscripts H and H', the carbon $2s$ orbitals and the various carbon p orbitals by X, Y, and Z, respectively.

$$\begin{vmatrix}
H_{SS}-E & H_{SX}-ES_{SX} & H_{SY}-ES_{SY} & H_{SZ}-ES_{SZ} & H_{SH}-ES_{SH} & H_{SH}-ES_{SH} & H_{SH}-ES_{SH} & H_{SH}-ES_{SH} \\
H_{SX}-ES_{SX} & H_{XX}-E & H_{XY}-ES_{XY} & H_{XZ}-ES_{XZ} & H_{XH}-ES_{XH} & H_{XH}-ES_{XH} & H_{XH}-ES_{XH} & H_{XH}-ES_{XH} \\
H_{SY}-ES_{SY} & H_{XY}-ES_{XY} & H_{YY}-E & H_{YZ}-ES_{YZ} & H_{YH}-ES_{YH} & H_{YH}-ES_{YH} & H_{YH}-ES_{YH} & H_{YH}-ES_{YH} \\
H_{SZ}-ES_{SZ} & H_{XS}-ES_{XS} & H_{SY}-ES_{SY} & H_{ZZ}-E & H_{ZH}-\mathrm{ES}_{ZH} & H_{ZH}-ES_{ZH} & H_{ZH}-ES_{ZH} & H_{ZH}-ES_{ZH} \\
H_{SH}-ES_{SH} & H_{XH}-ES_{XH} & H_{YH}-ES_{YH} & H_{ZH}-ES_{ZH} & H_{HH}-E & H_{HH'}-ES_{SH'} & H_{HH'}-ES_{HH} & H_{HH'}-ES_{HH'} \\
H_{SH}-ES_{SH} & H_{XH}-ES_{XH} & H_{YH}-ES_{HY} & H_{ZH}-ES_{ZH} & H_{HH'}-ES_{HH'} & H_{HH}-E & H_{HH'}-ES_{HH'} & H_{HH'}-ES_{HH'} \\
H_{SH}-ES_{SH} & H_{XH}-ES_{XH} & H_{YH}-ES_{YH} & H_{ZH}-ES_{ZH} & H_{HH'}-ES_{HH'} & H_{HH'}-ES_{HH'} & H_{HH}-E & H_{HH'}-ES_{HH'} \\
H_{SH}-ES_{SH} & H_{XH}-ES_{XH} & H_{YH}-ES_{SY} & H_{ZH}-ES_{ZH} & H_{HH'}-ES_{HH'} & H_{HH'}-ES_{HH'} & H_{HH'}-ES_{HH'} & H_{HH}-E
\end{vmatrix}$$

Hoffmann chose values for the one-centered integrals from experimental values in the literature as follows:

$$H_{ii}(C2p) = -11.4\,\text{eV}$$

$$H_{ii}(C2s) = -21.4\,\text{eV}$$

$$H_{ii}(H1s) = -13.6\,\text{eV}$$

The two-centered integrals are approximated as

$$H_{ij} = 0.5\text{K}(H_{ii} + H_{jj})S_{ij}$$

The value of K was taken as 1.75, and the values of the overlap integral S_{ij} were determined within the program. Long-range interactions are retained by keeping all of the two-centered integrals. One consequence of this is that molecular geometries are determined by the calculation.

The EHT method is easy to use, and the program is available for a number of computers.[7,8] In some versions,[8] the STOs are replaced by a linear combination of Gaussian orbitals. This basis set is described in more detail in Chapter 10. The EHT method starts from an input geometry either in the form of Cartesian coordinates or via some input graphics program. In the example for methane below, the input was generated by the graphics capabilities of the CAChe program, and the geometry was optimized with the MM2 force field. The chosen basis set was the STO-3G, which approximates the STO by the optimized linear combination of three Gaussian functions. The energy levels

Table 7.1. Eigenvalues for Methane as Determined by the EHT Method with an STO-3G Basis Set

1	2	3	4	5	6	7	8
−0.7796	−0.5699	−0.5699	−0.5699	0.3435	0.3438	0.3440	1.1568

Table 7.2. Eigenvectors for Methane by the EHT-STO-3G Method[a]

			1	2	3	4	5	6	7	8
1	1 C	2S	−0.5060	0.0000	0.0000	0.0000	0.0001	0.0000	0.0004	1.7445
2	1 C	2Px	−0.0000	−0.3054	0.4449	0.0526	−0.1448	−0.1389	1.2085	0.0005
3	1 C	2Py	0.0000	0.3967	0.2390	0.2818	−0.6062	−1.0468	−0.1930	0.0000
4	1 C	2Pz	0.0000	−0.2081	−0.1972	0.4602	−1.0544	0.6209	−0.0550	0.0000
5	2 H	1S	−0.2178	−0.2974	0.4333	0.0513	0.1375	0.1319	−1.1476	−0.7027
6	3 H	1S	−0.2177	0.4634	0.0749	0.2418	0.4970	0.8932	0.5557	−0.7020
7	4 H	1S	−0.2177	−0.0825	−0.0973	−0.5125	−1.1349	−0.0310	0.2537	−0.7018
8	5 H	1S	−0.2177	−0.2485	−0.4111	0.2194	0.5001	−0.9942	0.3391	−0.7020

[a]MO coefficients should be read in a descending order for each MO.

(eigenvalues) and MO coefficients (eigenfunctions) are given in Tables 7.1 and 7.2, respectively. The numbers reported here are somewhat different from those produced by using STO orbitals,[9] and the sign convention on the MO coefficients is reversed from that reported by Lowe (a matter of no consequence). A graphical representation of the first four filled MOs is given in Figure 7.1 in two views. The lowest MO has the same sign over all atoms. The next three MOs are degenerate, each holding a pair of electrons.

Molecular energies are given as total electronic energies in atomic units (hartrees, 1 hartree = 627.5 kcal/mol). Given a good starting point, molecular geometries are generally satisfactory, though strained systems are better handled by more sophisticated programs. Using this approach, Woodward and Hoffmann were able to draw satisfactory conclusions regarding the transition states for pericyclic reactions (Chapter 6) leading to the Woodward-Hoffmann rules.

The EHT method has been parameterized for the whole periodic table. This is not yet true for several of the most important semiempirical methods to be discussed in Chapter 9. As a consequence, the EHT method finds use mainly with organic metallic compounds and transition metal complexes and in providing guessed first structures for *ab initio* calculations.

7.3 The Self-Consistent Field or Hartree-Fock Method

If Hückel and extended Hückel methods suffer from failing to specifically account for electron and nuclear repulsions, how can these complicating interactions be included in a Hamiltonian such a way that a meaningful solution can be achieved? We saw that the Hückel methodology consisted of three parts. First, a Hamiltonian was written; second, a wave function was chosen; and third, Eq. 1.8 was solved for the energy levels and MO coefficients using the method of variations. No integrals were specifically evaluated, and

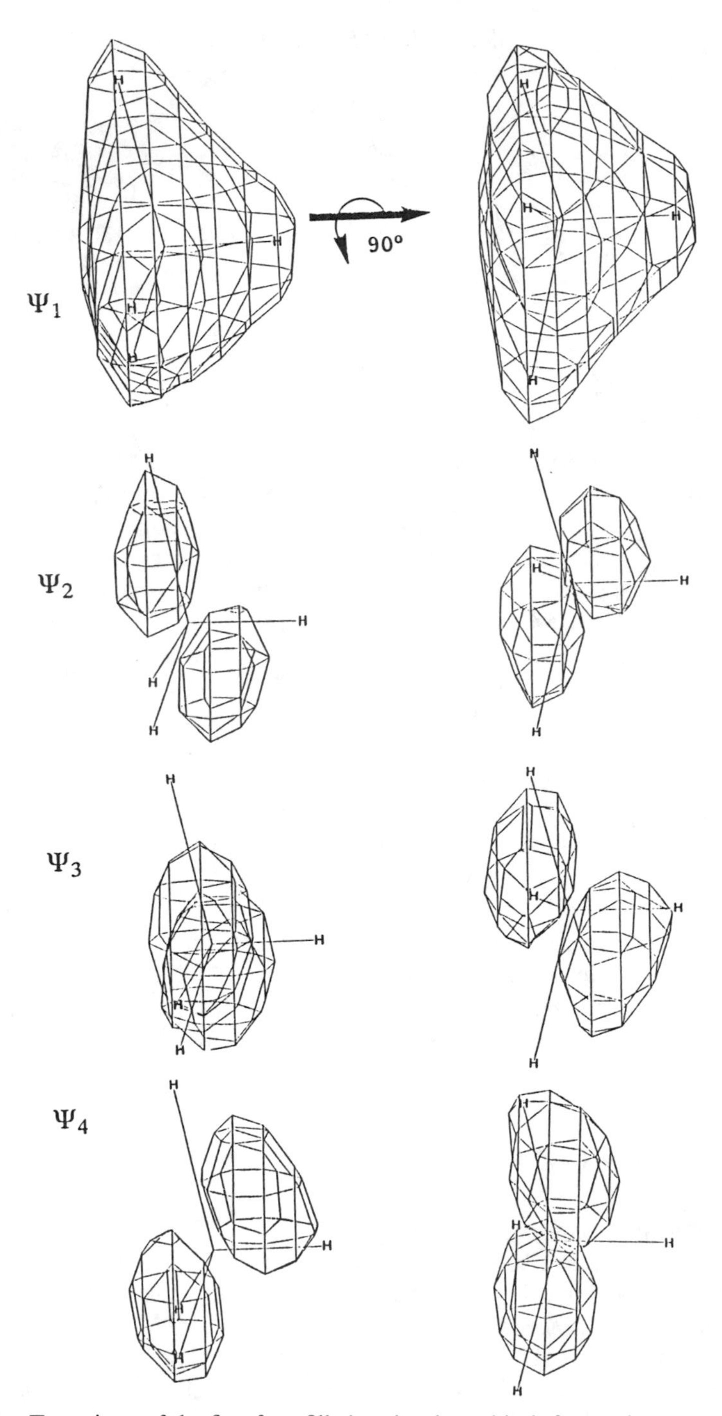

Figure 7.1. Two views of the first four filled molecular orbitals for methane calculated by the EHT method.

the mathematics was very deliberately simplified to make solution achievable. In principle, the same three steps could be employed using a more complex Hamiltonian with the various attractive and repulsive forces included, a more descriptive basis set of wave functions could be chosen, and the variation method could be again employed with the various integrals evaluated by modern computer techniques or assigned values based on experimental data and adjusted pragmatically to reproduce the properties of known trial molecules.

Let us address the wave function first. We assigned electrons to the HMOs with the statement that the spins must be paired, i.e., antiparallel. The reason for this now needs more explicit statement. The Pauli principle states that no two electrons in the system may have the same quantum numbers. For two electrons in the same orbital, all of the quantum numbers will be the same except for the spin, and it is this entity which must vary to obey the principle. Particles such as electrons, protons, and neutrons with half integer spins are called *fermions* and require description by antisymmetrical wave functions. Such wave functions will change signs when the electrons are interchanged. In contrast, photons, alpha particles, and others with integer spins require symmetrical wave functions. These particles are called *bosons*.

Using the customary notation (the symbols α for spin $+\frac{1}{2}$ and β for spin $-\frac{1}{2}$), we introduce here the concept of spin orbitals as a product of a spatial component ψ and a spin component α. Thus, for the two electrons in the ground state orbital of the hydrogen molecule, a product function stating the electronic configuration would be of the form $\psi_1(1)\alpha_1(1)\psi_1(2)\beta_1(2)$, where $\psi_1(1)\alpha(1)$ is the spin orbital for one electron and $\psi_1(2)\beta_1(2)$ is the spin orbital for the second. Next we must take into account that these electrons do not bear permanent labels and that they can be interchanged. If so, a second wave function, degenerate with the first, can be written, i.e., $\psi_1(2)\alpha_1(2)\psi_1(1)\beta_1(1)$. When two configurations such as these are degenerate, it is necessary to make a linear combination of both and renormalize, yielding

$$1/\sqrt{2}[\psi_1(1)\alpha_1\psi_2(2)\beta_1(2) \pm \psi_1(2)\alpha_1(2)\psi_1(1)\beta_1(1)]$$

The Pauli principle rules out the positive combination, since interchange of the spin labels would leave the signs the same. However, for the negative combination, interchange of spins multiplies the function by -1. This antisymmetric wave function meets the Pauli requirement:

$$\Psi_0 = 1/\sqrt{2}[\psi_1(1)\alpha_1\psi_2(2)\beta_1(2) \pm \psi_1(2)\alpha_1(2)\psi_1(1)\beta_1(1)] \tag{7.6}$$

The antisymmetric wave function can be expressed as a determinant format.

$$\Psi_0 = \frac{1}{\sqrt{2}}\begin{vmatrix} \psi_1(1)\alpha(1) & \psi_1(1)\beta(1) \\ \psi_1(2)\alpha(2) & \psi_1(2)\alpha(2) \end{vmatrix}$$

Such determinants are called Slater determinants, and they ensure the antisymmetry property because to exchange electrons is equivalent to exchange a row, an action which multiplies the determinant by -1. Recipes for setting up Slater determinants for larger electron sets are given in many quantum mechanics texts. Linear combinations of Slater determinants are sometimes used, with the coefficients for each being determined by the method of variations.

With regard to the Hamiltonian, let us start with the hydrogen molecule and generalize the result to multielectron-multiorbital molecules. We will use atomic units here to simplify the equations. The unit of length (the bohr) is equal to the most probable radius of the 1*s* electron in the hydrogen atom. The unit of charge is -1 a.u., which equals the charge on the electron. The unit of mass (1 a.u.) is the rest mass of the electron.

A diagram defining the system is given below. We return to our brief treatment of the hydrogen atom in Chapter 1 and apply the same thinking to the hydrogen molecule, with appropriate additions to express electrostatic attractions and repulsions between charged species, each of which bears a unit charge for the hydrogen case.

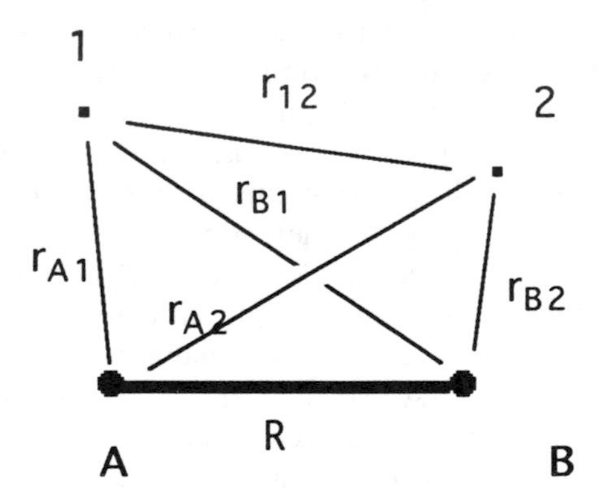

The Hamiltonian for the electronic energy is given by

$$\mathbf{H} = -\frac{1}{2}\nabla_1^2 - \frac{1}{2}\nabla_2^2 - \frac{1}{r_{A1}} - \frac{1}{r_{B1}} - \frac{1}{r_{A2}} - \frac{1}{r_{B2}} + \frac{1}{r_{12}} \tag{7.7}$$

If the total energy is wanted, then a term for nuclear repulsion must be added as $1/R$.

The above expression can be generalized to larger molecules as follows:

$$\mathbf{H} = -\frac{1}{2}\sum_{i=1}^{n}\nabla_i^2 - \sum_{\mu=1}^{N}\sum_{i=1}^{n}(\mathbf{Z}_\mu/r_{\mu i}) + \sum_{i=1}^{n-1}\sum_{j=i+1}^{n} 1/r_{ij} \tag{7.8}$$

where the i and j are indices for the n electrons and μ is an index for the nuclei N.

It is from this point of departure that the Hartree-Fock equations are developed. Readers wishing to follow the details at this point are referred to the text by Lowe.[9] In the following it will be assumed that we deal only with a closed-shell (paired electrons only) single determinantal form of the wave function. Application of the variational method leads to a result[9]:

$$\mathbf{F}\Psi_0 = E_0\Psi_0$$

The **F** is called the Fock operator, and the subscript zeros indicate that we are using the wave function for the ground state configuration. For a more complete rendition, the Fock operator for electron 1 is given by

$$\mathbf{F}(1) = -\frac{1}{2}\nabla_1^2 - \sum_{\mu}(Z_\mu/r_{\mu 1}) + \sum_{j=1}^{n}(2\mathbf{J}_j - \mathbf{K}_j) \tag{7.9}$$

The **J** operator is related to the attraction terms in Eq. 7.8, i.e.,

$$\mathbf{J}_j = \int \psi_j(2)(1/r_{12})\psi_j(2)\, d\tau(2) \tag{7.10}$$

the factor of two before the **J** in Eq. 7.9 reflects the double occupancy of the spatial orbitals.

The operator **K** finds its analog as an exchange operator:

$$\mathbf{K}_j\psi_i(1) = \int \psi_j(2)(1/r_{12})\psi_i(2)\, d\tau(2)\psi_j(1) \tag{7.11}$$

The differentiation between the Hamiltonian operator **H** above and the Fock operator **F** is that the latter is a function of Ψ and so cannot be known directly. We are trapped in a vicious circle, and iterative methods are called for. An initial guess at the wave functions is made using a method such as the extended Hückel or other semiempirical method. These orbitals are used to construct a Fock operator that is used to generate a new set of MOs. The process of refinement is repeated until a self-consistent set of orbitals and an energy minimum are reached. For closed-shell systems, this method is called the restricted Hartree-Fock (RHF) or self-consistent field (SCF) method. The description as "restricted" implies not only a single determinant wave function, but the placing of both α and β spins in the same orbital.

7.4 Configuration Interaction

The initial result of an RHF calculation is a set of MOs that describe the orbitals and orbital energies for the molecule. Electrons are placed in pairs starting with the lowest-energy orbital until all electron pairs have been placed.

This still leaves a set of unoccupied MOs called *virtual orbitals*. When electrons are excited, they move out of their lowest energy configuration into excited state configurations. These transitions correspond to lines in the electronic spectrum. One of the major deficiencies in the RHF method is a failure to correlate electron motion.

One way to correct for this deficiency is to deliberately combine the ground state wave function with a series of new configurations involving the promotion of electrons from their ground state orbitals into various virtual orbitals. These new configurations are written as Slater determinants and a lienar combination of ground and excited state determinants is then formed. If the entire range of possible excited state determinants is used, the computation is said to involve "full configuration interaction (CI)." As may be imagined, with a large molecule and many filled and virtual orbitals, the computational task rapidly becomes very large indeed. As a result, CI calculations are often limited to single and doubly excited states. The Slater determinants for the RHF ground state configuration and, say, the first two excited states are made into a linear combination with the appropriate coefficients:

$$\Psi_{\mathrm{CISD}} = c_1|\mathrm{Grd\ State}| + c_2|\mathrm{1st\ Excited\ State}| + c_3|\mathrm{2nd\ Excited\ State}|$$

The coefficients are then determined by the method of variations. This particular form of configuration interaction is called CI-singles, doubles (CISD). Higher orders of CI are only practical for small molecules or with very large, fast computers.

References

1. G. W. Wheland and D. E. Mann, *J. Chem. Phys.* **17**, 264 (1949).
2. A. Streitwieser, Jr., *Molecular Orbitals for Organic Chemists*, John Wiley & Sons, New York, 1961, p. 360.
3. F. Brogli and E. Heilbronner, *Theoret. Chim. Acta*, **26**, 289 (1972).
4. (a) R. McWeeny, *Coulson's Valence*, 3nd ed., Oxford University Press, Oxford, UK, 1979; (b) *Molecular Orbitals in Chemistry, Physics, and Biology*, P.-O. Löwdin and B. Pullman, eds., Academic Press, New York, 1964.
5. P.-O. Löwdin, *J. Chem. Phys.*, **18**, 365 (1950).
6. R. Hoffmann, *J. Chem. Phys.*, **39**, 1397 (1963).
7. Quantum Chemistry Program Exchange (QCPE), Creative Arts Building 181, 840 State Highway 46 Bypass, Bloomington, IN 47405.
8. CAChe Scientific, Beaverton, OR, phone 800-544-6634.
9. J. P. Lowe, *Quantum Chemistry*, 2nd ed., Academic Press, San Diego, CA, 1993.

CHAPTER

8

Molecular Modeling—Molecular Mechanics

8.1 The Nature of Force Fields

Each of the sciences progresses by modeling various aspects of their respective subjects. This is one way scientists can picture in their minds the way that nature works. Of course, one requirement is that the model must correspond to what experiment has shown us, and it is most valuable when it suggests new experiments which themselves will test the model. A history of the role of modeling in chemistry would probably start with an account of the ancient Greek concept of the atom as the smallest indivisible particle of matter, much like the grains of sand on a beach, or perhaps Dalton's atomic theory based on the behavior of gases and meteorological observations of water vapor in cloud formations.

Modern molecular modeling was born in the aftermath of the World War II when computers first began to become available. Computational molecular modeling has matured with the development of the computer, and two approaches have been employed: molecular mechanics and quantum mechanics. The first approach is the subject of this chapter.

Molecular mechanics has many facets in common with early molecular spectroscopy, which applied methods of classical mechanics to an understanding of vibrational and rotational spectra. Atoms are treated as hard spheres with fixed masses connected by springs with assigned bending and stretching force constants. Electrostatic attractions and repulsions are also included. The sum of the equations taking all these factors into account is called the "force field," and molecular mechanics calculations are sometimes called force field calculations. A number of force fields have been developed in recent years, each with a somewhat different set of basic assumptions and approximations. A simple force field is shown below and the various terms are described.[1]

$$E_{\text{total}} = E_{\text{stretch}} + E_{\text{bend}} + E_{\text{torsion}} + E_{\text{nonbonded}} \tag{8.1}$$

When two atoms (A and B) approach each other, resulting in the formation of the diatomic molecule A – B, the energy of the system follows the familiar Morse potential energy plot. As A and B approach from infinity, the electrons around each nucleus become polarized, leading to an attractive London or dispersion force. The energy continues to decrease until the equilibrium bond length r_0 is reached. At shorter bond distances, repulsion between electron charge clouds becomes increasingly important

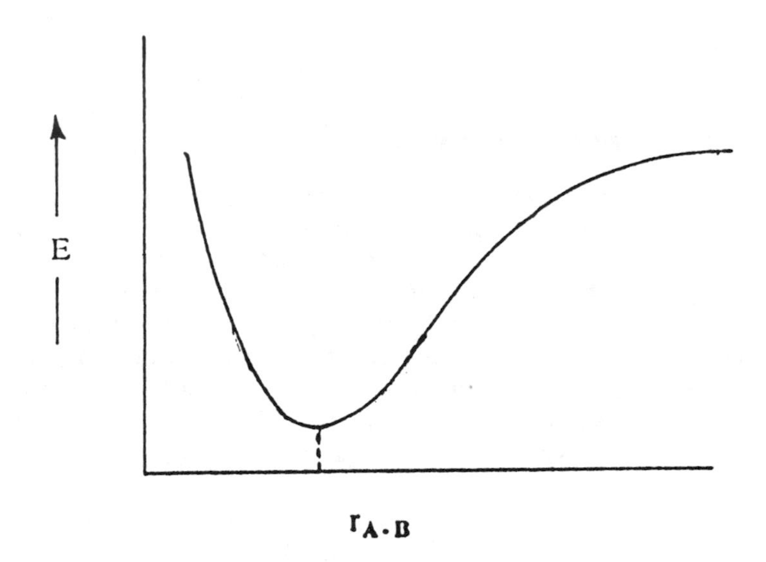

(van der Waals repulsion) and the energy quickly mounts.

Inspection of the curve above shows that in the region around the equilibrium bond distance, the curve closely approximates a parabola. Bond stretching forces in this region may be represented by a Hooke's law type of interaction, thus:

$$E_{\text{str}} = k_{\text{str}} \frac{(r - r_0)^2}{2} \tag{8.2}$$

Similarly, the three atom-bending interactions can be approximated by

H
C Θ
H

$$E_{\mathrm{bnd}} = k_{\mathrm{bnd}} \frac{(\theta - \theta_0)^2}{2} \tag{8.3}$$

Early force fields attempted to account for hydrocarbon energies without taking into consideration the variations due to torsion angle changes. It has been known for a number of decades now that the heat capacity of ethane, butane and cyclohexane could not be accounted for in a satisfactory way without including terms that reflected the effects of rotation about carbon–carbon single bonds. Furthermore, it was evident from the thermal data that rotation about these bonds was not "free."

Modeling data for ethane from the program PCMODEL[2] are given below.

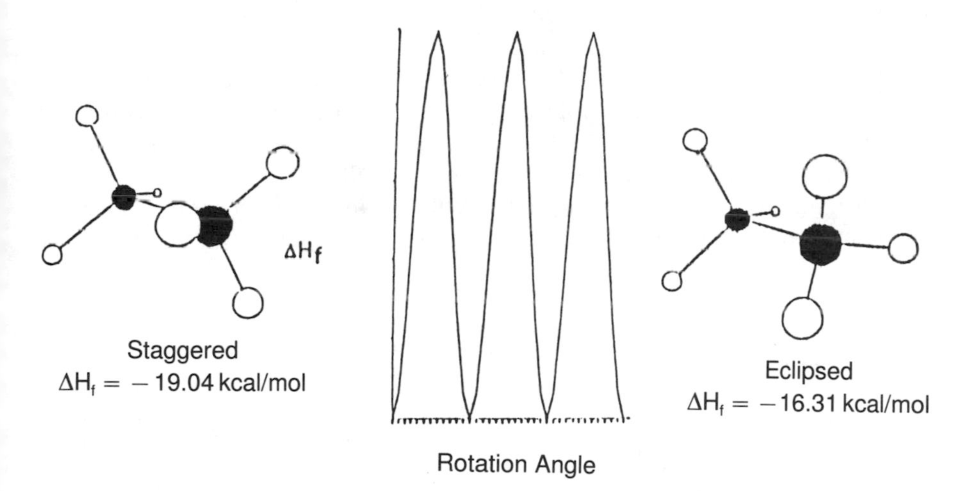

As can be seen, the eclipsed form of ethane is calculated to be 2.7 kcal/mol higher in energy than the staggered form. In this instance, the difference is reflected in the calculated heats of formation. We'll discuss this term below. A plot of the change in energy is also shown, starting from a staggered conformation and rotating through 360°. These energy swings come about as a result of repulsive electronic interactions that occur when the C—H bonds are close together in space, i.e., an eclipsing interaction.

The final term (Eq. 8.1) in our simplified force field is a nonbonding interaction reflecting the London dispersion forces, which are attractive at long distances, and the counterbalancing van der Waals repulsion, which set in when the distance between atoms approaches the sum of their van der Waals radii. This term has been cast in a number of different forms by the developers of various force fields. Many of these start with a potential function first developed many years ago by the English physical chemist Leonard-Jones. The Leonard-Jones potential has the form below:

$$E_{\text{nonbonded}} = \varepsilon\left[\left(\frac{r_0}{r}\right)^{12} - 2\left(\frac{r_0}{r}\right)^{6}\right] \tag{8.4}$$

where ε is the energy value at the minimum in the curve coinciding with r_0.

The force field described above is not meant to represent any one of the many existing force fields available to today's researchers. It is meant, rather, to give some appreciation of the complexities of the problem and to suggest a general format. Table 8.1 summarizes the potential functions for four popular molecular mechanics programs. Each particular force field defines the mechanical model to be used in computing a molecular structure and its accompanying energy. The process starts with the force field. Next, the chemist must provide the molecular structure which is to serve as the starting point for the calculations. The atom types must be specified and a connection table provided. This can be done either through the input of a computer file generated by as crude a method as taking Cartesian coordinates off a framework or space-filling molecular model, or by means of a computer graphical drawing program. The latter method is employed by the popular programs such as PCMODEL and CAChe, as well as others.

Once an input structure has been entered into the program, a set of parameters describing the molecular geometry is computed, i.e., bond lengths,

Table 8.1. Partial Force Fields for Various Programs[a]

Stretch	Bend	Torsion
MM2[1,4]		
$k^*(\Delta r^2 + \Delta r^3)$	$k^*(\Delta\theta^2 + \Delta\theta^6)$	$V_1(1 + \cos\omega) + V_2(1 - \cos 2\omega) + V_3(1 + \cos 3\omega)$
MM3[3]		
$k^*(\Delta r^2 + \Delta r^3 + \Delta r^4)$	$k^*(\Delta\theta^2 + \Delta\theta^3 + \Delta\theta^4 + \Delta\theta^5 + \Delta\theta^6)$	$V_1(1 + \cos\omega) + V_2(1 - \cos 2\omega) + V_3(1 + \cos 3\omega)$
Amber[5]		
$k^*(\Delta r^2)$	$k^*(\Delta\theta^2)$	$V_n/2(1 + \cos(n\phi - \gamma))$
Charm[6]		
$k^*(\Delta r^2)$	$k^*(\Delta\theta^2)$	$V_n/2(1 + \cos(n\phi - \gamma))$

[a]K. E. Gilbert, 207th National Meeting, American Chemical Society, 1994.

and bond and torsion angles. These values are then fed into the terms of the force field equation and a steric energy is calculated. Generally this energy will be expressed in kcal/mol and is the sum of all the potential energy terms contained in the force field. The steric energy is a value specific for a given force field. Such numbers cannot be used to compare values calculated by other programs. Furthermore, as a generalization, such steric energies should not be used to compare the relative stabilities of different molecules, though they may be used to compare different conformations of the same molecule.

The next step in the process is to alter the structure in a systematic way to minimize the steric energy. This process will carry the molecule to an energy minimum somewhere on the molecular potential energy surface. Although a variety of mathematical approaches to the minimization process have been developed, the Newton-Raphson method has served as a model for many of these and is still much used today. The basic concepts of this method may be illustrated graphically as in Figure 8.1, with a one-dimensional plot of energy E_x as a fourth-order polynomial function of a bond stretching variable x. The equation and its first and second derivatives are given below. These are plotted for various values of x in the figure. The first derivative determines the gradient of the energy with respect to molecular geometry at any point on the E_x curve. The program is written in such a fashion as to carry the structure down the energy path or until a minimum is reached at $f'(x) = 0$ (points 1 and 3 above). Since $f'(x)$ is also zero at a maximum (2), it is usual to include a measure of the curvature $f''(x)$ as part of the calculation. This function is also plotted below. As can be seen, when $f'(x)$ crosses the baseline at E_x equals zero, $f''(x)$ has positive values for the minima at points 1 and 3 but a negative value for the maximum at point 2. This description conveys the basic concepts of the Newton-Raphson method of minimization, which is still much used today as well as serving as a starting point for many modern minimization techniques.

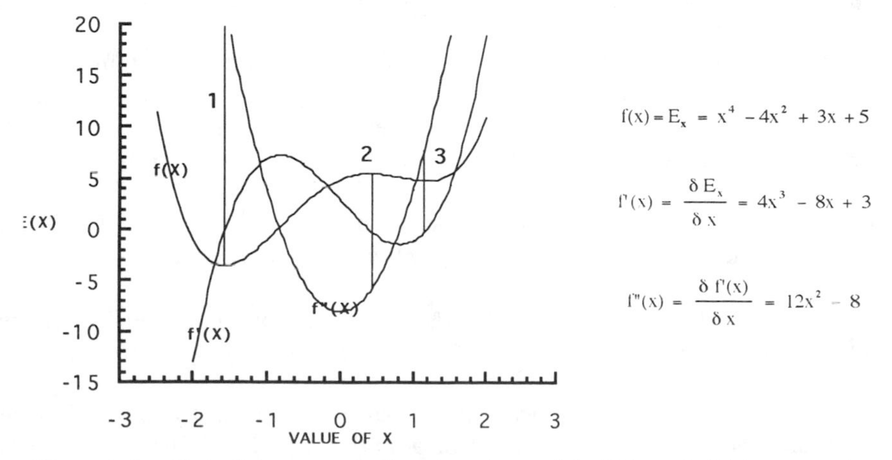

Figure 8.1. Plot of the example function along with $f'(x)$ and $f''(x)$ as noted.

Procedures of this type are applied as well to quantum mechanical structure-energy minimization programs.

At this point it might be wise to enter a caveat for the reader, and it is simply this: *A molecular mechanics calculation will take the input structure you provide to its nearest local minimum on the potential energy surface.* This does not guarantee that one has calculated the most stable form of the molecule. Indeed, sometimes the computation may have difficulties in recognizing the input structure and may carry the calculation to conclusion on a nonsense structure. Those who do these calculations need to apply their chemical intuition to make sure that the answer provided by the program makes chemical good sense. The process of finding a global minimum is considerably more complex than the process just described, and this will be discussed briefly later in the chapter.

Some programs provide only the steric energy as output. This rather limits their utility. Most, however, provide also a computed heat of formation and a term that measures the strain energy in the molecule. The latter value is of a more controversial nature, as it presupposes a knowledge of the heat of formation for the same molecule in a strain-free configuration. Each program treats this assumption in a different way. Strain energies are meaningful within any given program but will differ from program to program. The same should not be true for the heat of formation. Calculation of the heat of formation requires the application of a group or bond increment value applied to the steric energy. The heat of formation is defined as the energy required or released when the molecule is formed from its respective elements at 25°. For instance, for carbon dioxide and for water, the heat of formation is just the heat of combustion, since by definition the value for a free element is zero.

$$C + O_2 \longrightarrow CO_2 \qquad \Delta H = -94.1 \text{ kcal/mol}$$

$$H_2 + 1/2\, O_2 \longrightarrow H_2O\,(l) \qquad \Delta H = -68.4 \text{ kcal/mol}$$

As a generalization, the heat of formation for a series of homologous alkanes is -5.1 kcal/mol per methylene group and -10.1 kcal/mol for a methyl. Since heats of formation can be determined experimentally, this is a particularly useful term to know, as experimental versus computed values can be used as a test of the quality of a given force field. Table 8.2 offers computed and experimental values for several molecules as determined by PCMODEL and the currently latest force field developed by the Allinger group at the University of Georgia (MM3).[3]

It should be pointed out that PCMODEL uses a modified version of the Allinger MM2 force field and predates the version of MM3 used in these calculations.

Another insight into the heat of formation is given by the following definition. The heat of formation of compound A equals the sum of the heats of formation of the products minus the heat of combustion of A. Using ethane

Table 8.2. Experimental and Calculated Heats of Formation (kcal/mol) for Representative Organic Molecules

Compound	Experimental	PCMODEL	MM3[a]
Ethane	−20.24	−19.08	−20.05
Pentane	−35.00	−36.27	−35.17
Cyclobutane	6.78	5.96	6.29
Cyclohexane (chair)	−29.43	−28.53	−29.96
trans-Decalin	−43.54	−43.76	−43.60
Bicyclo[2.2.0] hexane	29.90	26.33	30.66
Norbornane	−12.42	−12.84	−12.04

[a] MM3 data quoted from N. L. Allinger, Y. H. Yuh, and J.-H. Lii, *J. Am. Chem. Soc.*, **111**, 8551 (1989).

as an example,

$$CH_3 - CH_3 \quad + \; 7/2 \; O_2 \longrightarrow 2 \; CO_2 \quad + \quad 3 \; H_2O \qquad -373.0 \text{ kcal/mol}$$

$$(2 \times -94.1) + (3 \times -68.4) = -393.4 \text{ kcal/mol}$$

$$-393.4 - (-373.0) = -20.4 \text{ kcal/mol calc'd vs } -20.2 \text{ experimental}$$

8.2 The Problem with π-Systems

It is perfectly possible to include in the force field a set of parameters for sp^2 hybridized carbons found in olefins and aromatic systems. The atom types for the carbons will have to reflect the lengths of the carbon–carbon double bond (ca. 1.34 Å for olefins and 1.40 Å in benzene).

The solution to this problem has been to include a quantum mechanical π calculation in the molecular mechanics program. One popular system is the Variable Electronegativity SCF (VESCF) method, which is based on the work of Brown and Heffernen.[7] As a comparison with the results on the benzyl cation given previously, the charge densities as calculated by the VESCF program embedded in PCMODEL are given below:

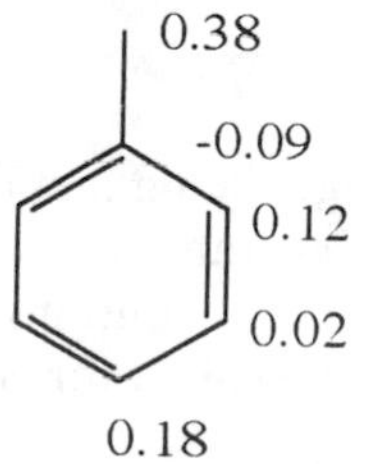

0.38
-0.09
0.12
0.02
0.18

Once the input structure has been entered into the program, a calculation is carried out and the bond lengths are adjusted according to the π bond order. The force field calculation is then carried through enough cycles to produce an energy-minimized structure. Given this new geometry, the π calculation is repeated. The whole process is recycled until no significant improvement is found.

The structures for 1,4-pentadiene and *trans*-1,3-pentadiene are shown below.

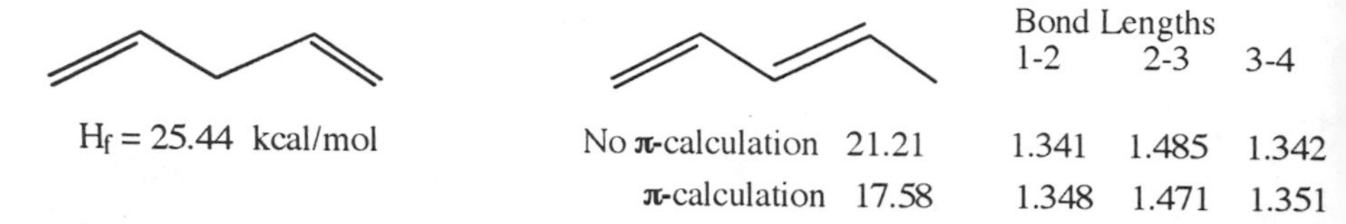

Each structure was minimized by PCMODEL. The experimental heats of formation are 25.3 and 18.1 kcal/mol, respectively. For the *trans*-1,3-diene, the heats of formation and the bond lengths of interest are calculated for minimizations carried out with and without the π interactions. As can be seen, the conjugated double bond is slightly longer than its ethylenic value, just as the sp^2–sp^2 single bond is shortened.

A similar set of calculations on benzene and naphthalene are included below. When the π calculation is not included, benzene is treated as 1,3,5-cyclohexadiene with alternating single and double bonds. The experimental heats of formation for benzene and naphthalene are 19.8 and 36.1 kcal/mol, respectively. When no π calculation is carried out, the bond lengths in both cases correspond to alternating ethylenic and sp^2–sp^2 single bond lengths. With the π calculation, benzene bond lengths are all equal at 1.40 Å, and naphthalene shows the expected bond length alteration of 1.380 and 1.423 Å between carbons 1-2 and 2-3.

No π calculation
H_f = 52.90 kcal/mol
SCF π calculation
H_f = 19.25 kcal/mol

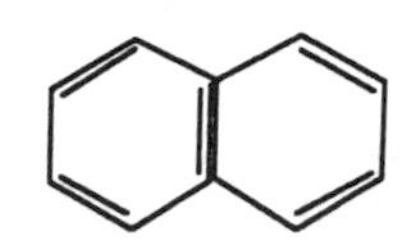

No π calculation
H_f = 89.88 kcal/mol
SCF π calculation
H_f = 35.21 kcal/mol

One of the interesting challenges being met by current research in molecular modeling is the attempt to model transition states.[8] PCMODEL includes, as no doubt do many of the existing programs, model transition states for a number of pericyclic reactions as well as for Sn2 displacement reactions. The user is asked to fill in certain details, usually bond orders for the transition state, and append his own particular structure to the framework provided. The present example is the Diels-Alder addition of ethylene to butadiene. The forming bonds were each assigned bond orders of 0.3. As is evident, both the ethylene and the butadiene hydrogens are distorted from their normal configurations. The new bonds are only starting to form, but already the old bond lengths are being significantly altered. If one wished, it would be possible to carry out the calculations at a variety of bond orders and make a movie of the overall reaction as it progressed. At some point in the bond-making process, the structure passes through an energy maximum corresponding to the transition state:

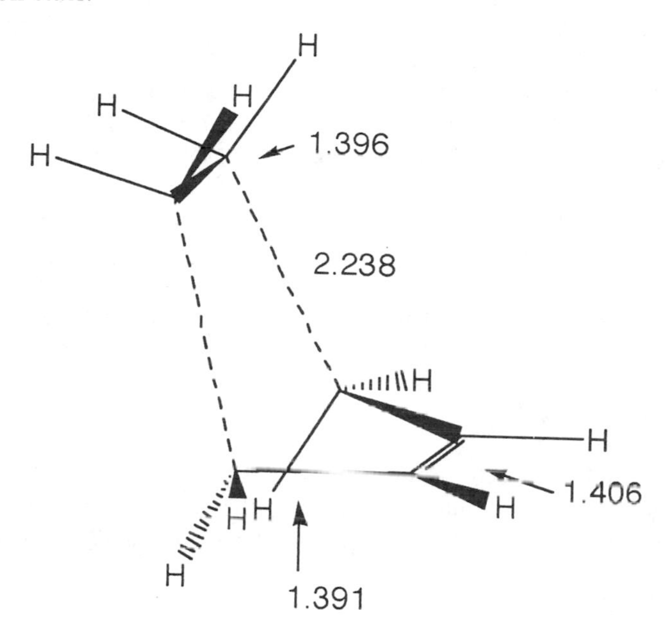

8.3 The Rearrangement of 2,7-Dimethylocta-2,6-diene(1)—A Case Study[9]

In a current popular textbook, a problem is given in which **1** is isomerized in phosphoric acid to 1,1-dimethyl-2-isopropenylcyclopentane (**4**). The experimental evidence supporting this contention was based on an old study in the literature where the point is made that the structure was considered probable, but not certain. The student is asked to offer a mechanism for this product with the object of establishing that this possible extension of the Saytzeff rule to this situation fails.

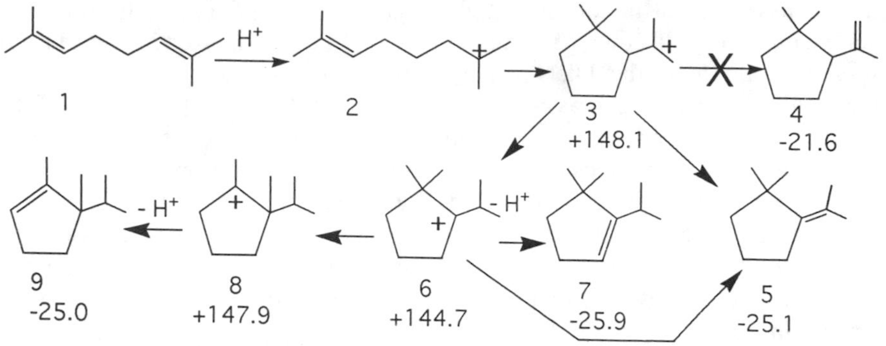

However, the expected E1 product from **3** would be expected to be the more highly substituted product **5** in keeping with expectations of the Saytzeff rule. A possible reason for the reaction favoring **4** might be the steric interaction of the isopropylidene methyls in **5** with the adjacent gem dimethyls. As shown above, calculations of the heats of formation of **4** and **5** with PCMODEL, however, rule against this possibility.

Another possible route for the reaction is rearrangement of the ion **3** to the more stable cation **6**. PCMODEL has been parameterized to handle alkyl carbocations and shown to give reasonable results when compared to the few examples of such ions where heats of formation are known. The carbocation **6** can deprotonate, giving the olefin **7**, or could rearrange to the less stable ion **8**, which would lead to **9** as a product. In fact, a search of the literature provides a more recent study of the rearrangement in which **7** and **9** are shown to be the major products along with a small amount of **5**. This is consistent with the computations above regarding the relative olefin ($7 > 5 \approx 9 \gg 4$) and cation ($6 \gg 8 \approx 3$) stabilities.

8.4 The Conformational Multiple Minimum Problem[10]

Saunders et al.[10b] have calculated that if one limited the available conformations to only *gauche* or *anti* torsion angles, the 17-carbon cycloheptadecane would have 4,782,969 possible conformations. If the minimization of a given input structure carried that structure to the nearest downhill minimum on the potential energy surface, then hypothetically one could input all the reasonable conformations and minimize each to determine the global minimum structure. This would prove a rather daunting task for cycloheptadecane. Such an approach might work for a relatively small molecule or for structures containing a great deal of rigidity. But a more systematic approach, allowing

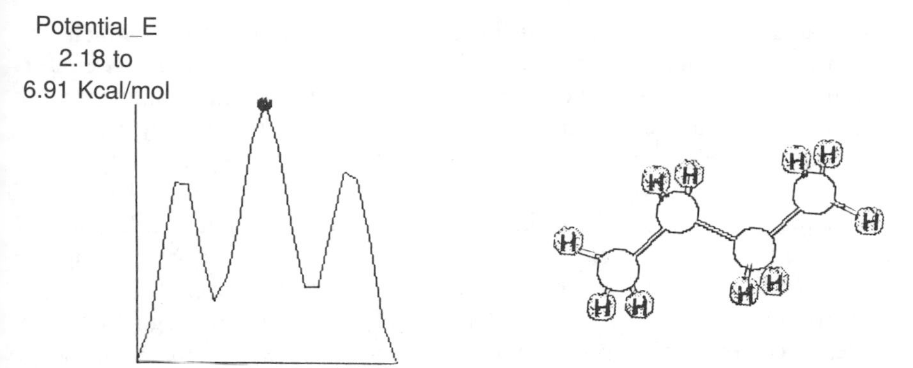

Figure 8.2. Plot of rotation about the C2—C3 bond of butane showing the relative energies of the *anti*, *gauche*, and eclipsing conformations as a function of torsion angle.

the computer to search the surface, is needed for large, flexible molecules, and a number of procedures have been developed to carry out global searches. Furthermore, many of these programs will collect all the structures within a given energy window, arrange them in order of increasing energy, and calculate the Boltzman distribution of conformers for a given temperature.

Two examples of limited search techniques are illustrated in Figures 8.2 and 8.3 as applied to the conformational mobility of butane. The molecule was first minimized to a local minimum. The central bond was then selected and dihedral driver program was instructed to carry out a series of minimizations during which the central C—C bond was rotated through a set sequence of

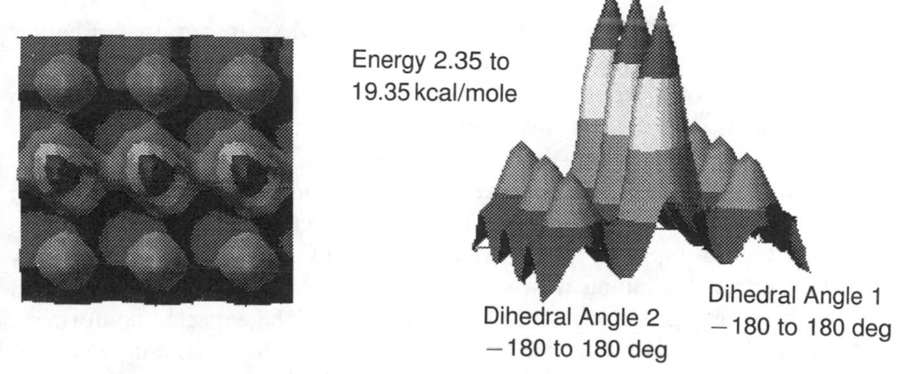

Figure 8.3. Two views of the energy surface generated by rotating two bonds in the butane molecule.

torsion angles for this bond. In this instance the CAChe program,[11] which uses the MM2 force field, was employed. The low-energy form is the *anti* conformation at the extremes of the energy plot. Starting from this conformer, the molecule is carried over an energy barrier to a local minimum corresponding to the *gauche* conformer. The two *gauche* conformers are separated by a high energy hill corresponding to the eclipsing of the two methyl groups. This would be a simple example of a more general technique of searching called a grid search. Grid searches are referred to as "deterministic" because the investigator sets the conditions which are to be varied throughout the search. In Figure 8.3, two torsion angles (CH_3—CH_2 and CH_2—CH_2) are varied through 360° each. The result is a three-dimensional map of the energy surface where steric energy (kcal/mol) is plotted against the two torsion angles.

The highest point on the surface corresponds to the eclipsed conformer with the methyl hydrogens interfering with each other. The number of possible combinations of bond rotation rapidly increases as molecular size increases. Then too, there is the problem of how to handle systems with one or more rings as part of the structure. Grid searches clearly have limited applicability.

Several modes of searching the potential energy surface have been devised based on randomized movements of bonds or atoms.[9] In a statistical search method, the initially minimized structure is altered by randomly moving a specified set of atoms in Cartesian space and each structure is minimized. A similar process may be applied to the rotation of bond torsion angles. A mixed method, varying both bonds and angles, is often the method of choice for small to medium-sized molecules. For ring compounds the ring is broken, and the atoms and angles are moved. If the result can be ring closed without introducing too much distortion, the minimization is carried out. Otherwise, the structure is discarded. In each of the above methods, the program summarizes those structures which have been saved within a set energy window, and a Boltzman distribution of the various conformers at 25° is calculated.

Application of the global search program GMMX[2] to a rather complex conformational problem is illustrated below.[12] The molecule (E)-cyclodec-5-enone is capable of a series of very complex twisting motions. Consideration of data on related structures in the literature had suggested those shown on the next page as distinct possibilities for conformational minima. The search method employed here was the mixed statistical method which searched on both bonds and atom coordinates. The initial structure was generated using PCMODEL and input to the GMMX program. Only the lowest-energy conformer and those falling within a 3.5 kcal/mol window were kept. As can be seen, the search program did, in fact, find each of the expected conformers. However, the forms with energies of 12.1 and 12.4 kcal/mol make up 56% and 35%, respectively, of the conformer mix. The NMR proton spectrum was consistent with this result.

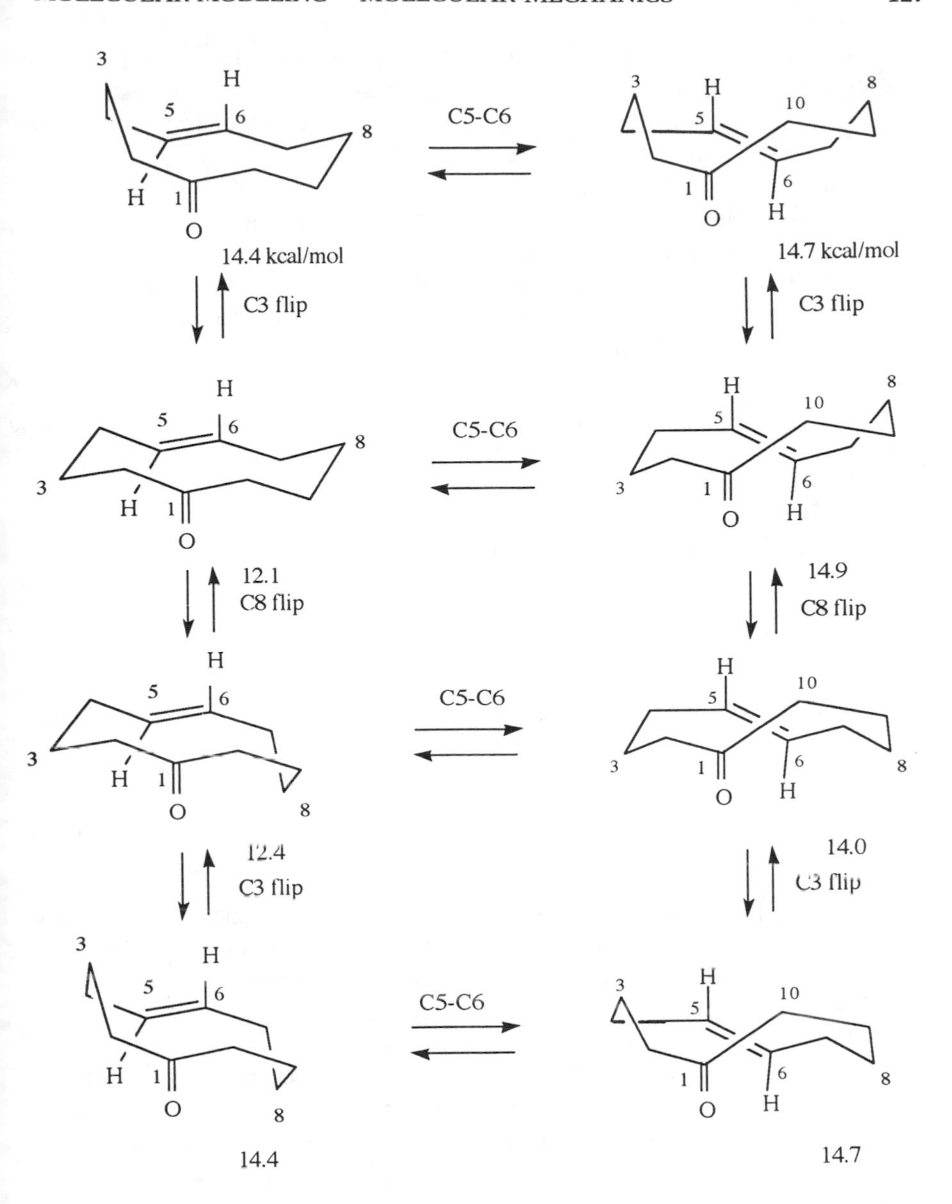

8.5 Molecular Dynamics[13]

Molecular dynamics (MD) is a third method of searching a potential energy surface. MD is assuming importance for studying large molecules of biological importance as well as modeling liquid structure and solvent effects. The application of this method to small organic molecules has been limited to this point in time (1995), and few examples of this method are to be found in the literature of organic chemistry. Since this method does not require separate minimizations of the molecule at each step, it is possible that the use of MD will grow as programs become more available.

The ideas embodied in MD may be considered as a combination of the molecular mechanics force field with Newton's laws of motion. If we ignore the effects of translational and molecular rotational kinetic energy, thereby confining our considerations to stretching, bending, and internal rotations as measures of the vibrational motions of atoms in a molecule, the average kinetic energy (KE) can be written as

$$\overline{KE} = \sum_{i}^{i=N} \frac{1}{2} m_i v_i^2 = NkT \tag{8.5}$$

where N is (3 × number of atoms −6).

The force on atom i at any time t is given by

$$F_i = m_i a_i(t) = m_i \frac{\partial^2 r_i(t)}{\partial t^2} \tag{8.6}$$

where m is the mass of atom i, a is the acceleration of atom i at time t, and the force on atom i at time t is given by

$$F_i = -\frac{\partial}{\partial r_i} V(r_1, r_2, \ldots, r_N) \tag{8.7}$$

where N is now the number of atoms in the molecule. The energy given to the molecule is set by the investigator and is usually requested as a temperature. From this temperature a Gaussian distribution of velocities can be computed for each atom. The energy for each atom is selected from this distribution. The amplitude of atomic motion is related to the square root of the temperature. The molecule is set into motion, and since it is isolated in a vacuum it can never lose energy by collisions with another molecule. However, vibrational and rotational energy will be redistributed throughout the molecule as time goes on. The relationship between temperature and atomic velocities (v_i) at a given time (t_i) is

$$T = \frac{1}{(3N - 6)k} \sum_{i=1}^{N} m_i |v_i|^2 \qquad i = 1, N$$

where k is the Boltzman constant. At specified times (usually each femtosecond; 1 fs $= 10^{-15}$ s), Newton's equations of motion are integrated and the molecular energy and structure at the time are recorded as a trajectory. For small to medium-sized organic molecules, trajectories are usually over a span of 1 to 3

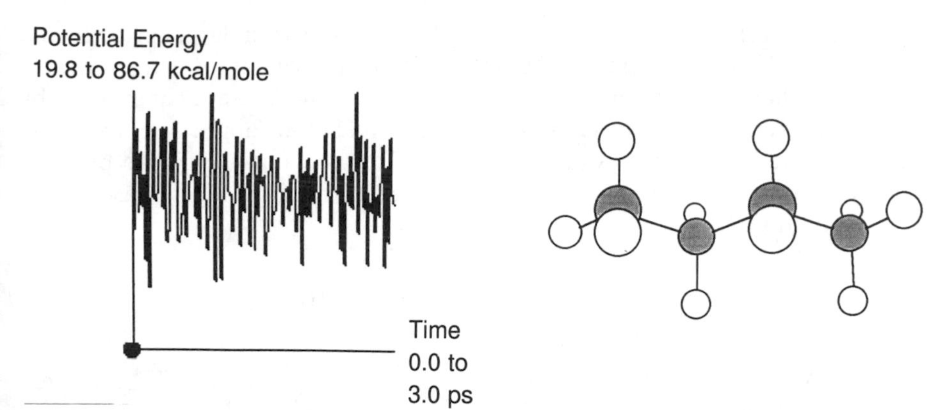

Figure 8.4. MD trajectory for rotation about the central bond in butane as acquired by the program CAChe.

or 4 picoseconds (1 ps = 10^{-12} s). However, for proteins, polymers, or cases where whole ensembles of solvent and subject molecules are studied, the time span may cover several hundred picoseconds, requiring a massive data storage capability.

Figure 8.4 shows a trajectory for the dynamic simulation of motions in butane. The total time period is 3 ps with sampling every 1 fs at a temperature of 600°. To prevent an unduly long trajectory file, only every twentieth point was saved to the file. When the cursor is moved along the trajectory, one can follow the changing conformations of the molecule on the right. The software (CAChe) will also do this for the observer in the form of a movie. The highest point corresponds to an eclipsing of the two methyls, as in Figure 8.3. Examination along the minima yields both *gauche* and *anti* conformations. The energies are steric energies given in kcal/mol. Comparison of various conformations is allowed here, since they all pertain to the same molecule. No computation of the conformer populations is given in calculation.

8.6 Monte Carlo Searches

In contrast to the deterministic methods described previously, the Monte Carlo method of searching is dependent on probability theory. A major difference from MD methods is that successive configurations are not generated by solving the Newtonian equations of motion. Instead, a starting configuration is subjected to a random series of changes in atomic positions. Each new configuration is minimized, and the potential energy is compared to that of the preceding structure. If the new energy is less, the structure is kept. If the new structure is greater in energy, it is discarded except in the case where the term $\exp(-\Delta V/kT)$ is less than a randomly chosen number between 0 and 1.

Monte Carlo methodology is important in studying large protein and polymer molecules that are not readily studied by molecular mechanics. Monte Carlo methods are often used to compute thermodynamic quantities by statistical thermodynamics. An interesting example of the application of Monte Carlo methods is a recent study by Cao et al.[14] which addresses the problem of the tautomeric equilibria of 3-hydroxpyrazole both in the gas phase and in water solutions.

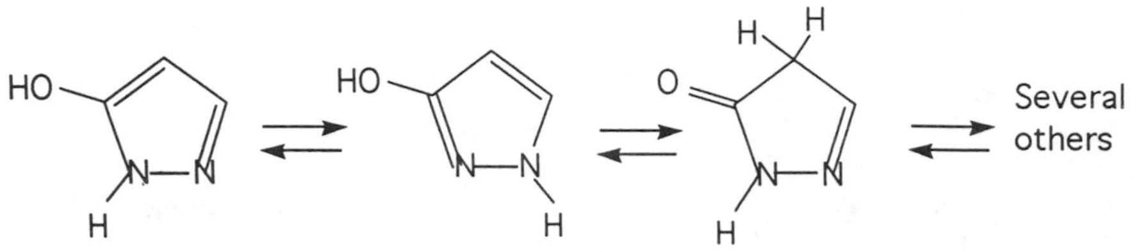

8.7 A Final Caveat

New molecular mechanics programs are coming onto the market almost monthly. Some are very expensive or require the use of very large computers. Others will run on desktop-size computers and work stations. Each program has its own set of limitations. One program is parameterized primarily to run very large molecules such as proteins. Another handles small molecules quite well but is not able to cope with conjugated systems such as butadiene or styrene. The buyer needs to investigate carefully the virtues of each program to see if it will meet his needs and computer capabilities.

Problems

The author used the program PCMODEL operating on a Macintosh IIci. There is also a PC version of this program as well as a number of competing programs which should lead to the same conclusions.

1. Minimize the methylcyclohexane with the methyl group equatorial and axial. If we assume that the entropies of these two forms are approximately equal, we can make use of the relationship below to calculate the equilibrium constant at room temperature. Calculate the value of K and the

   $$\Delta H_f = -RT \log K$$

   percents of axial and equatorial methylcyclohexane.

2. Rotational conformers of acetaldehyde include the two forms shown below. Calculate the heats of formation for each and draw conclusions as to which is the more stable form.

3. Calculate the heats of formation of butadiene in the two conformations shown below.

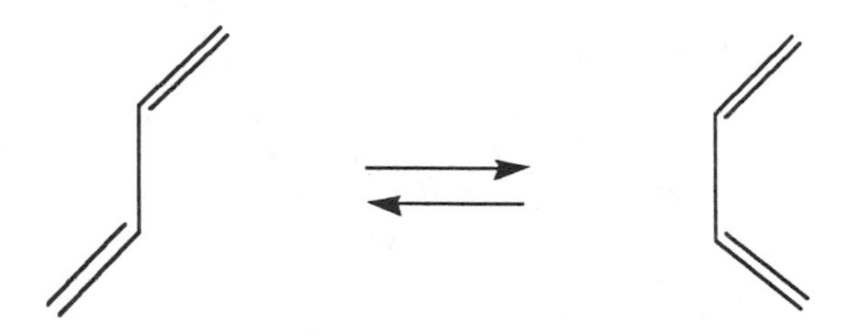

4. Most molecular mechanics programs will handle hydrogen bonds. The two conformations shown below are the most likely for the compound shown. Calculate the stabilities of each and compare with the results for methylcyclohexane. Be sure to tell the program to look for hydrogen bonds in each case.

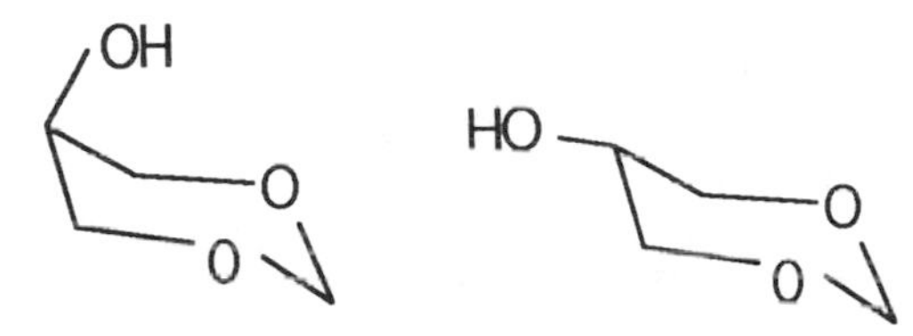

5. A rather dramatic demonstration of ring strain is given by the heats of formation and strain energies calculated for the two bicyclo[2.2.1]heptenes shown below. The first is an example of Bredt's rule. Calculate these values for each molecule.

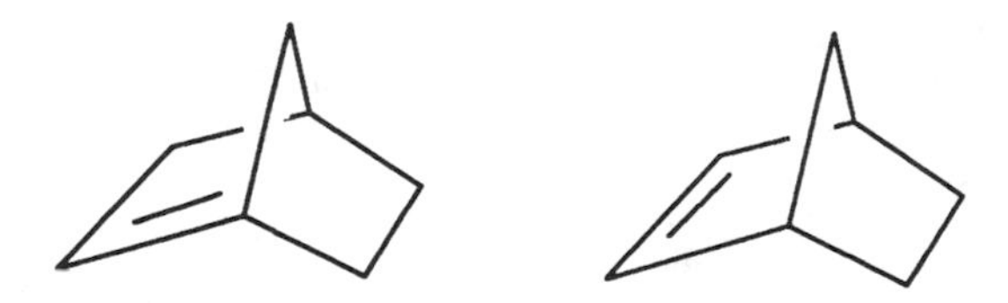

References

1. Additional details on the subject of molecular mechanics can be found in H. Burket and N. L. Allinger, *Molecular Mechanics*, American Chemical Society, Washington, DC, 1982; and T. Clark, *A Handbook of Computational Chemistry*, John Wiley & Sons, New York, 1985.

2. Serena Software, Bloomington, IN. See also J. J. Gajewski, K. E. Gilbert, and J. McKelvey, MMX, an Enhanced Version of MM2, *Advances in Molecular Modelling*, Vol. 2, JAI Press, 1990.

3. N. L. Allinger, Y. H. Yuh, and J.-H. Lii, *J. Am. Chem. Soc.*, **114**, 1 (1992).

4. Quantum Chemistry Program Exchange (QCPE), University of Indiana, Bloomington, IN 47405.

5. S. J. Weiner, P. A. Kollman, D. A. Case, U. Chandra Singh, C. Ghio, S. Profeta, and P. Weiner, *J. Am. Chem. Soc.*, **106**, 765 (1984).

6. B. R. Brooks, R. E. Bruccoleri, B. D. Olafson, D. J. States, S. Swaminathan, and M. Karplus, *J. Comp. Chem.*, **4**, 187 (1983).

7. R. D. Brown and N. L. Heffernen, *Aust. J. Chem.*, **12**, 319 (1959).

8. J. E. Eksterowicz and K. N. Houk, *Chem. Rev.*, **93**, 24439 (1993).

9. M. G. Reinecke and W. B. Smith, *J. Chem. Ed.*, **72**, 541 (1995).

10. For reviews of global search techniques see (a) A. E. Howard and P. A. Kollman, *J. Med. Chem.*, **31**, 8127 (1988); and (b) M. Saunders, K. N. Houk, Y.-D. Wu, W. C. Still, M. Upton, G. Chang, and W. C. Guida, *J. Am. Chem. Soc.*, **112**, 1 (1990).

11. CAChe Scientific, P.O. Box 4003, Beaverton, OR 97076.

12. Y. Chu, D. Colclough, J. B. White, and W. B. Smith, *Magn. Reson. Chem.* **31**, 937 (1993).

13. T. P. Lybrand in *Reviews in Computational Chemistry*, K. B. Lipkowitz and D. B. Boyd, eds., VCH Publishers, New York, Vol. 2, 1990, Chap. 8.

14. M. Cao, B. J. Teppen, D. M. Miller, J. Pranata, and L. Shafer, *J. Phys. Chem.*, **98**, 11353 (1994).

CHAPTER

9

Molecular Modeling—Semiempirical Methods

The Hückel and extended Hückel methods are examples of semiempirical methods, in that no integrals are ever evaluated. An appeal is made to experimental numbers when integral values are required. The parameterization is then tested against a limited set of molecules to ensure its accuracy. Like numerical iteration, this process is a continuing one. New versions of semiempirical programs appear almost yearly. These contain new computational capabilities and new parameters as improvements come to light.

Shortly after the introduction of the EHT method, Pople and co-workers at Carnegie-Mellon University[1] came forth with the earliest of a new breed of semiempirical methods. As exemplified in Chapter 7, the secular determinant for methane is an 8×8 determinant containing a number of Coulomb and overlap terms of the form H_{ij} and S_{ij}. We also saw that by using a perturbation treatment, it was possible to drop explicit consideration of the overlap integral, resulting in a scaling of the eigenvalues. The secular determinant for the Hartree-Fock method can be presented by analogy to Eq. 2.6 as

$$|F - ES| = 0 \tag{9.1}$$

Given the earlier observations on methane, it is not hard to imagine the consequences of moving to larger molecules. The number of off-diagonal one-electron integrals H_{ij} and S_{ij} rapidly becomes very large and computationally burdensome. To handle this task, the decision was made to set the overlap integrals between different atomic orbitals to zero. With this assumption the determinant reduces to

$$|F - E| = 0 \tag{9.2}$$

It should be remembered that the Fock matrix here is the sum of the usual one-electron and two-electron positions. The latter present a particularly difficult hurdle, since a number of integrals of the form

$$\iint \psi_i(1)\psi_j(2)(1/r_{12})\psi_k(1)\psi_1(2)\, d\tau(1)\, d\tau(2)$$

are encountered. Fortunately, the overlap assumpton sets these integrals to zero except in the case where $i = j$ and $k = l$. This assumption led to the designation CNDO (complete neglect of differential overlap). An additional assumption was that the off-diagonal resonance integral H_{ij} could be made proportional to the overlap integral, i.e., although the overlap matrix disappears out of Eq. 9.1, values are still assigned to certain S_{ij} to allow evaluation of the H_{ij} and for later computation of the charge distribution (Mulliken population analysis). A somewhat later upgrade in the parameterization was forthcoming under the title CNDO/2. Although the CNDO methods did introduce electron–electron repulsions, they do not handle the question of the interactions between electrons with parallel or antiparallel spins, especially when the electrons are on the same atom.

As a correction, differential overlap between electrons on the same atom as covered by one-electron integrals was reintroduced. This improved semi-empirical method, called the intermediate neglect of differential overlap (INDO), has been expanded by Zinder at the University of Florida to provide better geometries than in the earliest versions. One version (INDO/S) has been parameterized specifically for the prediction of molecular spectra. A collection of CNDO/1, CNDO/2, and two new versions of INDO—one for geometry and one for spectral calculations—was made under the general term ZINDO. The ZINDO programs have been incorporated into the CAChe software.[2] An example is included here.

9.1 INDO—A Case Study

Watson and Sun at Texas Christian University have prepared a series of metal complexes derived from naphthoquinone disulfide. The nickel (zero) complex is shown, along with the LUMO, which is centered entirely on the four sulfurs. This complex is diamagnetic, but the −1 complex is paramagnetic:

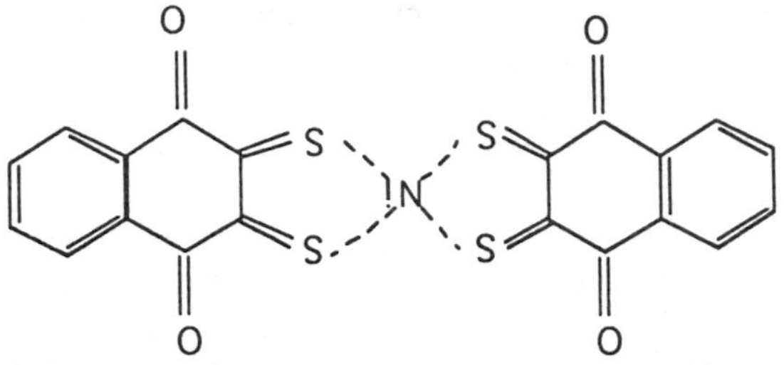

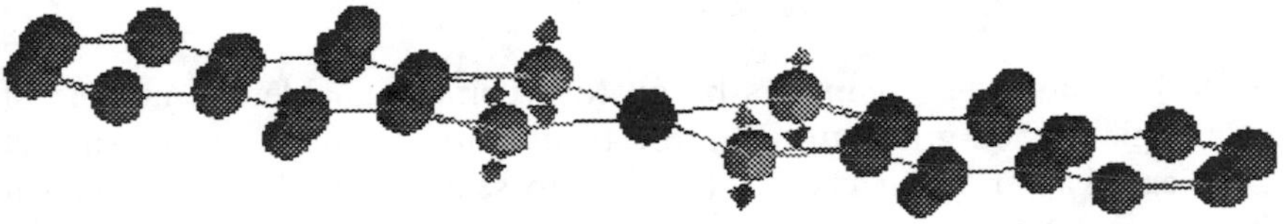

Transitions between various electronic states are responsible for the absorption of electromagnetic energy in the visible and ultraviolet portions of the spectrum. Since the various filled and virtual orbitals are computed during the course of all electron computations, these methods offer the possibility of computing electronic spectra. INDO uses single-determinant orbitals in an RHF SCF method. However, to compute the spectral transitions, excited state orbitals as well as filled orbitals must be computed. Configuration interaction is required to adequately represent these excited states. INDO, as incorporated into the ZINDO package, provides a rapid method for obtaining geometries that are considered to be fairly accurate, along with the necessary levels of CI to allow a reasonable approximation of the electronic spectrum. Electronic energy levels actually have a considerable amount of fine structure as a result of the effects of imposed vibrational and rotational energies. Since electronic transitions between two energy levels may include changes within these vibrational-rotational sublevels, appreciable line broadening may be observed. Because of the difficulties of including CI at sufficiently high levels, Hartree-Fock theory is known to give only approximate values for electronic transitions. Research papers reporting spectral calculations often state a correction factor to be applied to the computed transition frequencies. The electronic spectrum for the nickel compound above is given here. The experimental values were found to fall at 296, 340, and 496 nm, with intensities closely replicating those shown. Lines in the near infrared were found at 635, 825, and 863 nm, corresponding to the broad band in the computed spectrum. No corrections have been applied to the computed frequencies in the spectrum shown below. These results required a few minutes only on a Macintosh Centris 650 computer.

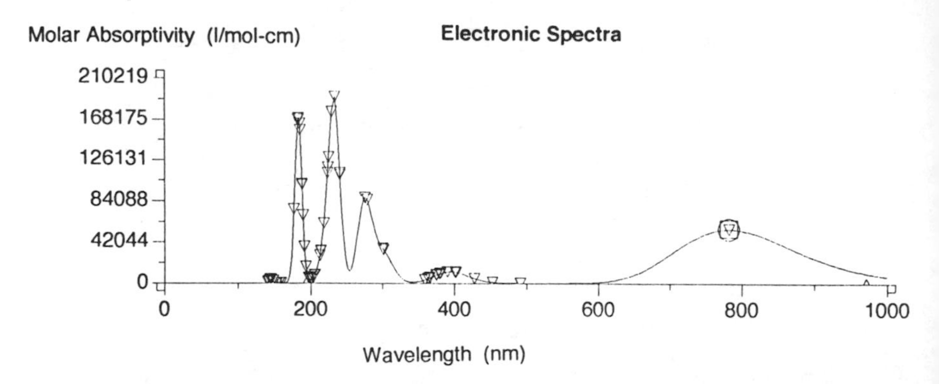

As larger and faster computers began to become available, the interest of the Carnegie-Mellon group turned more to *ab initio* calculations, the subject of the next chapter. Their efforts in regard to semiempirical methods ended with the NDDO method (neglect of diatomic differential overlap), in which the only overlap integrals excluded were those on different atoms. This reduction in the original CNDO assumption greatly increased the computational burden but allowed for the inclusion of lone pair repulsions, which had been a failure of the earlier methods.

9.2 AMPAC and MOPAC

In the late 1960s and early 1970s, interest in the development of semiempirical methods grew in the laboratories of M. J. S. Dewar at the University of Texas. Dewar had been instrumental in the development of the PMO method and began to feel the need for a computationally more exact method of applying MO theory. Given the extant computational difficulties of *ab initio* methods, he and his group set out to supply programs that would allow chemists to rapidly calculate molecular geometries, energies, dipole moments, and spectral properties and to explore reaction pathways with an optimum accuracy for the given computer size and demands of time. Today it is possible to run detailed calculations on large molecules in minutes or hours on desktop computers and even more rapidly on mainframe computers and workstations. There is still argument as to the virtues of the approach used by the Dewar group, as opposed to the more rigorous methods, but there is no question that many chemists do get quite useful results with a minimum of expenditure in time and money from the semiempirical approach. Over the years, a series of complex programs have been put into single packages, allowing the chemist to calculate molecular energies, geometries, thermodynamic properties, electronic and vibrational-rotational spectra, and solid state and polymer properties. Upon the retirement of Professor Dewar, various colleagues have assumed responsibility for these programs and their further development. The program AMPAC is currently handled by Professor Andrew Holder,[3] and Dr. J. J. P. Stewart has assumed the directorship of the MOPAC program.[4] Versions of AMPAC and MOPAC are available through the Quantum Chemistry Exchange Program as well as a number of other sources. These programs have many points of similarity, since both originated within the Dewar research group. The following remarks will apply specifically to MOPAC version 6 and MOPAC93, since these are the versions most familiar to the author.

Early Hamiltonians developed by the Dewar group (MINDO, MINDO/3, and MNDO) may be considered as extending Pople's earlier work with differing treatments of the various kinds of resonance and overlap integrals. One starts by providing a trial molecular structure. The program establishes a linear combination of Slater type s and p orbitals for this structure and calculates a trial density matrix defined as

$$P_{AB} = 2\left(\sum_{i}^{\text{occ}} c_{Ai}c_{Bi}\right) \tag{9.3}$$

The reader will recognize the definition of charge density from HMO theory here. The trial wave functions and the trial density matrix are used to define a crude Fock matrix, and the iterative SCF calculation is initiated. The cycles repeat through new eigenfunctions to better density matrices to yet better Fock matrices, etc. Once an energy-minimized geometry has been achieved, the program carries out additional tasks, such as CI calculations, charge densities, dipole moments, and spectral calculations.

Current versions of MOPAC allow the user to choose between four different Hamiltonians (listed in chronological order of appearance: MINDO/3, MNDO, AM1, and PM3). MINDO/3 was the first Dewar program specifically parameterized to give heats of formation, a quantity that many organic chemists find more useful than total electronic energies.

MNDO (modified neglect of diatomic overlap) is a semiempirically parameterized version of NDDO concepts. MNDO did not give good results in describing the hydrogen bond. AM1 (Austin Model 1)[5] is a more recent update with further parameterization that does successfully treat hydrogen bonds. In the process, the number of parameters required to describe each atom effectively doubled. PM3 (Parametric Method 3) continued the shifting of parameterizations based on spectral data to molecular data. An automated optimization routine was introduced which utilized a large data bank of molecular information and could derive parameters for several elements at once. PM3 has been reviewed by Stewart.[6] At the present writing, there are not enough comparative studies in the literature to allow a meaningful statement about the relative virtues of PM3 versus AM1. The various MOPAC Hamiltonians are parameterized for a different number of elements. The elements of greatest interest to organic chemists are common to all. Individual versions should be checked in the accompanying manual to be sure the elements of interest are to be found in the method you choose.

9.3 MOPAC Data Files

Calculations with either AMPAC or MOPAC start with the generation of an input file in the MOPAC format. Unless specified to the contrary, the calculation will be an RHF SCF calculation using the MNDO Hamiltonian. If one uses a program with a graphical input as the front end (CAChe), the input structure can be generated on the computer screen and minimization initiated without a detailed examination of the input file. The author uses PCMODEL to generate his MOPAC files, since he uses MOPAC versions available for either Macintosh, Windows, or a mainframe VAX for computation. Most MOPAC versions will output a structure file of the minimized structure. PCMODEL can read these files. Hypothetically, one could generate files by reading angles and distances off a framework model and generating the input file in this fashion.

The input file consists of three introductory lines followed by a data file specifying an input geometry. The file flows as follows:

First line: Consists of a series of keywords describing the desired Hamiltonian and any special conditions for the calculation or special types of calculations desired, i.e., saddle calculation, force calculation, frequency calculation, or calculations in solvents of differing dielectric constant (MOPAC93). These keywords are described in the manual that accompanies the program, and a few examples are given subsequently.

Second line: Contains identifying comments describing the calculation.

Third line: Is to be left blank.

Fourth and following lines: There will be one line for each atom, which will indicate the atom type, the distance to the preceding attached atom, the bond angle to the preceding two atoms, the torsion (dihedral) angle to the preceding three atoms, instructions as to which parameters are to be minimized, and a table of atom connectivities. The input file must terminate with a blank line, as shown below.

In Chapter 2 the HMO calculation of cyclobutadiene was discussed; the result was a square planar molecule with a triplet ground state—an unsatis-

```
Input File:
 1   PM3
 2   SQUARE CYCLOBUTADIENE from PCMODEL
 3
 4   C   0.000000  0     0.000000  0     0.000000  0  0  0  0
 5   C   1.436693  1     0.000000  0     0.000000  0  1  0  0
 6   C   1.436680  1    90.000534  1     0.000000  0  2  1  0
 7   C   1.436693  1    90.000526  1     0.000000  1  3  2  1
 8   H   1.101661  1   135.000275  1   180.000000  1  1  2  3
 9   H   1.101661  1   135.000275  1   180.000000  1  4  3  2
10   H   1.101661  1   134.999710  1   180.000000  1  2  3  1
11   H   1.101661  1   134.999725  1   180.000000  1  3  2  1
12
```

factory answer resulting from the HMO theory applied at the simplest level. In the following example, PCMODEL was used to generate a MOPAC input file for a square planar cyclobutadiene to see if a PM3 calculation would correct the situation. The line numbers are given as a convenience in describing the input file. The columns of 0s and 1s tell the program which variables are to be optimized. The PM3 keyword gives the required Hamiltonian to be used. The file was converted to a text file, and executed on a Macintosh IIci. An abbreviated output file is given below.

Output File:

```
*                 MOPAC:  VERSION  6.00                CALC'D.  31-JAN-95
*  T=             -A TIME OF100000.0  SECONDS REQUESTED
*  DUMP=N         -RESTART FILE WRITTEN EVERY   3600.0 SECONDS
*  PM3            -THE PM3 HAMILTONIAN TO BE USED

PM3
cyclobutadiene
                              BOND
ATOM       CHEMICAL          LENGTH        BOND ANGLE    TWIST ANGLE
NUMBER      SYMBOL        (ANGSTROMS)      (DEGREES)      (DEGREES)
(I)                          NA:I          NB:NA:I       NC:NB:NA:I     NA  NB  NC
1              C
2              C           1.43669 *                                     1
3              C           1.43669 *       90.00000 *                    2   1
4              C           1.43669 *       90.00000 *      0.00000 *     3   2   1
5              H           1.10166 *      135.00001 *   -180.00000 *     3   2   1
6              H           1.10166 *      135.00001 *   -180.00000 *     4   3   2
7              H           1.10166 *      135.00005 *   -180.00000 *     1   2   3
8              H           1.10166 *      135.00003 *   -180.00000 *     2   3   1
CYCLE:       1 TIME:    2.00 TIME LEFT:    99987.0         81.609 HEAT: 101.4168
                                            GRAD.:
CYCLE:       2 TIME:    3.00 TIME LEFT:    99984.0         69.422 HEAT:  99.88409
                                            GRAD.:
CYCLE:       3 TIME:    1.00 TIME LEFT:    99983.0         21.998 HEAT:  99.19867
                                            GRAD.:
TEST ON GRADIENT SATISFIED
          HOWEVER, A COMPONENT OF GRADIENT IS LARGER THAN 1.00
CYCLE:       4 TIME:    1.00 TIME LEFT:    99982.0          3.845 HEAT:  99.11226
TEST ON GRADIENT SATISFIED
PETERS TEST SATISFIED
                              PM3     CALCULATION
                                                             VERSION  6.00
                                                             31-JAN-95
                FINAL HEAT OF FORMATION =        99.10704 KCAL
                TOTAL ENERGY            =      -531.59398 EV
                ELECTRONIC ENERGY       =     -1659.62118 EV
                CORE-CORE REPULSION     =      1128.02720 EV
                IONIZATION POTENTIAL    =         8.69494
                NO. OF FILLED LEVELS    =        10
                MOLECULAR WEIGHT        =        52.076
                SCF CALCULATIONS        =        10
```

ATOM NUMBER (I)	CHEMICAL SYMBOL	BOND LENGTH (ANGSTROMS) NA:I	BOND ANGLE (DEGREES) NB:NA:I	TWIST ANGLE (DEGREES) NC:NB:NA:I	NA	NB	NC
1	C						
2	C	1.54476 *			1		
3	C	1.34823 *	90.00866 *		2	1	
4	C	1.54478 *	90.00746 *	−0.01606 *	3	2	1
5	H	1.07892 *	138.25002 *	179.96887 *	3	2	1
6	H	1.07862 *	131.39472 *	179.50946 *	4	3	2
7	H	1.07859 *	131.38455 *	−179.97376 *	1	2	3
8	H	1.07890 *	138.28038 *	179.98519 *	2	3	1

EIGENVALUES

−38.11703 −24.80663 −22.47489 −17.97074 −17.29421 −14.60325 −13.76100
−12.41772 −11.68441 −8.69494 −0.06051 2.04490 2.89739 3.05282
3.75075 3.86331 4.54770 4.95825 5.38445 6.35060

NET ATOMIC CHARGES AND DIPOLE CONTRIBUTIONS

ATOM NO.	TYPE	CHARGE	ATOM ELECTRON DENSITY
1	C	−0.1291	4.1291
2	C	−0.1304	4.1304
3	C	−0.1303	4.1303
4	C	−0.1290	4.1290
5	H	0.1297	0.8703
6	H	0.1297	0.8703
7	H	0.1297	0.8703
8	H	0.1297	0.8703

DIPOLE	X	Y	Z	TOTAL
POINT-CHG.	−0.005	0.000	0.005	0.007
HYBRID	0.005	0.000	0.002	0.005
SUM	0.000	0.000	0.007	0.007

ATOMIC ORBITAL ELECTRON POPULATIONS

1.22187 0.94450 0.96272 1.00000 1.22216 0.94497 0.96328 1.00001
1.22216 0.94495 0.96324 0.99999 1.22186 0.94443 0.96270 0.99999
0.87034 0.87026 0.87025 0.87033

Examination of the output shows that the molecule has been minimized to a rectangular structure of alternating single and double bonds with a singlet electronic configuration. The basis set contains 20 terms (4×4 carbon orbitals $+4$ hydrogen orbitals), leading to 20 new molecular orbitals, of which 10 are filled. A ball-and-stick representation generated by converting the structure output by the MOPAC file with the program Chem3D[7] is shown in Figure 9.1, as are the HOMOs and LUMOs.[8]

It is of interest to compare the heats of formation of a series of related compounds with both the experimentally available values and the molecular mechanics (PCMODEL) values. This is done in Table 9.1. Going from 1-butene to butadiene results in one-half the energy change found in passing from cyclobutene to cyclobutadiene. The enormous increase in energy is due,

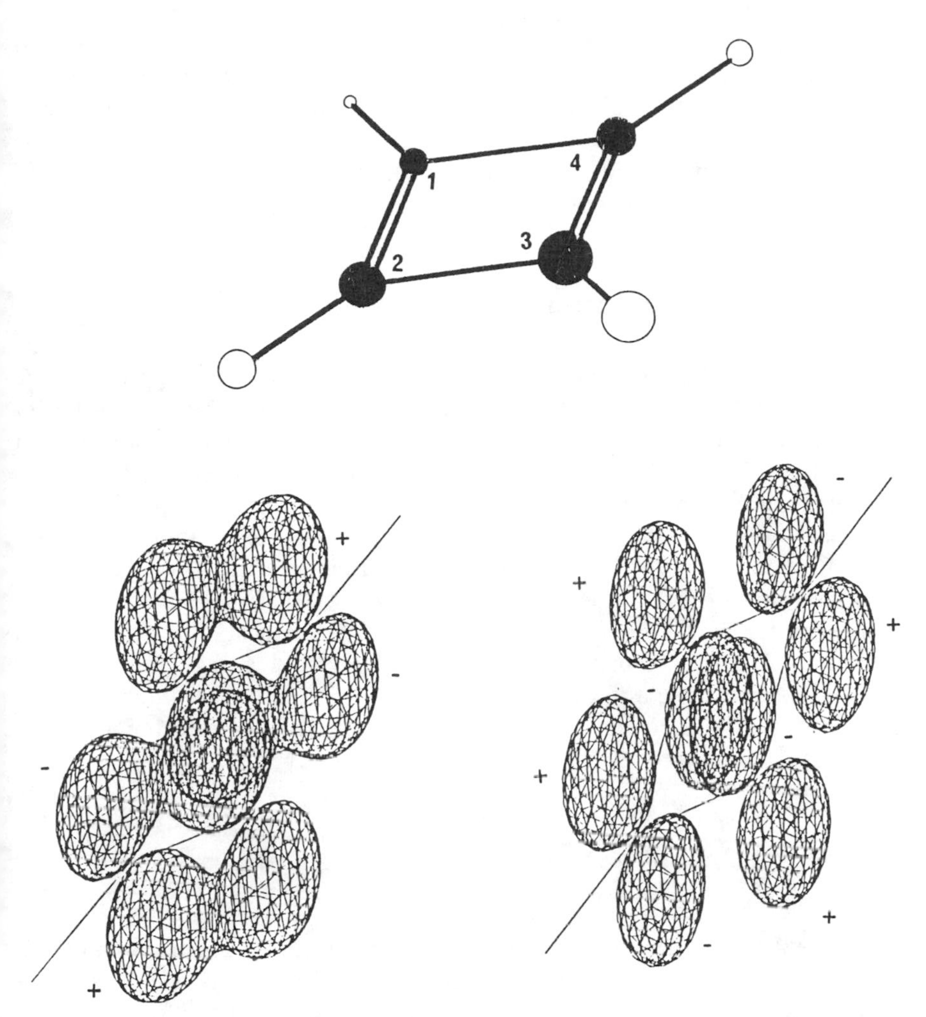

Figure 9.1. Ball-and-stick structure for cyclobutadiene above with the AM1-generated HOMO and LUMO displayed below.

Table 9.1 Experimental and Calculated Heats of Formation (kcal/mol) for a Related Series of Butenes

	H_f(Exp)	H_f(PCMODEL)	H_f(PM3)	ΔH_f
1-Butene	−0.2	−0.0	1.8	
1,3-Butadiene	26.0	25.6	31.0	29.2
Cyclobutene	37.5	37.7	37.7	
Cyclobutadiene	—	90.2	99.1	61.4

no doubt, to bond angle strain, and accounts for the great difficulties experienced in preparing the latter compound. Considering the entirely different methods of computation and parameterization, the agreement between the molecular mechanics and the semiempirical results may be all that can be expected at this time. Had we used the AM1 Hamiltonian, the H_f values would have been somewhat different but not necessarily in better agreement.

9.4 Mullikan Population Analysis

In Chapters 2 and 5, we found the electron charge density to be a useful gauge of physical and chemical properties. As calculated for the π electrons in HMO theory, all charges were concentrated on atoms in the π system. In all electron calculations, the situation becomes more complicated as substantial amounts of charge density are found in the regions of orbital overlap. Several schemes have been devised to cope with this problem, but there is an almost universal incorporation of the Mullikan population analysis scheme in available standard programs—both semiempirical and *ab initio*. We saw the format for the initial density matrix in Eq. 9.3 and noted the similarity to the Coulson definition of charge density. After the geometry and energy minimization process has converged and the eigenvalues and eigenvectors are known, a final density matrix is calculated. We can define an electron density function $\rho(\mathbf{r})$ such that

$$\rho(\mathbf{r}) = \sum_{\mathrm{A}}^{N} \sum_{\mathrm{B}}^{N} P_{\mathrm{AB}} \psi_{\mathrm{A}} \psi_{\mathrm{B}} \tau \tag{9.4}$$

where the ψ values are normalized basis functions from a linear combination of N-basis functions, and n is the total number of valence electrons. The probability of finding an electron in a small volume element $d\mathbf{r}$ is just $\rho(\mathbf{r})\, d\mathbf{r}$. Integrating over all of space then gives

$$\int \rho(\mathbf{r})\, d\tau = \sum_{\mathrm{A}}^{N} \sum_{\mathrm{B}}^{N} P_{\mathrm{AB}} \int \psi_{\mathrm{A}} \psi_{\mathrm{B}}\, d\tau = \sum_{\mathrm{A}}^{N} \sum_{\mathrm{B}}^{N} P_{\mathrm{AB}} S_{\mathrm{AB}} = n \tag{9.5}$$

When A = B, the overlap integral $S_{\mathrm{AB}} = 1$ and Eq. 9.5 reduces to

$$\sum_{\mathrm{A}}^{N} P_{\mathrm{AA}} + 2 \sum_{\mathrm{A}}^{N} \sum_{\mathrm{B}}^{N} P_{\mathrm{AB}} = n \tag{9.6}$$

The value of P_{AA} is the density on the atomic center A. Given several choices, Mullikan shared the density in the overlap region AB equally between the two centers. The gross population associated with ψ_A is given by

$$q_A = P_{AA} + \sum_{B \neq A} P_{AB} S_{AB} \tag{9.7}$$

and for the N-basis functions

$$\sum_{A}^{N} q_A = N \tag{9.8}$$

The atomic charge then is just

$$Q_A = Z_A q_A \tag{9.10}$$

The individual charges and atom-charge densities for cyclobutadiene as calculated by a Mullikan population analysis are given in the output data above. Note that the sum of the latter is 20—just the number of valence electrons used in the calculation.

9.5 Electrostatic Potentials

In commenting on the Mullikan population analysis as a charge partitioning scheme, Hehre et al.[9] point out that, although it is the most widely used method for attributing charge densities, it is also the most criticized. Several other methods have developed over the years. Although each of these has its proponents, perhaps the single most useful method for organic chemists is that of electrostatic potentials. The reasons for this statement will become evident shortly.

In classical electrostatic terms, the electrical potential V_p at a given point P is defined as the work required to bring a unit positive charge from infinity up to that point. If the field at the point consists of contributions from a number of point charges q_i, then the potential in MKS units is given by the equation

$$V_p = \frac{1}{4\pi\varepsilon_0} \sum_{i=1}^{n} \frac{q_i}{(r_i - r)} \tag{9.11}$$

where r_i and r are the positions of the point charges and the test charge, respectively. The signs of the various terms in the summation may be positive or negative, depending on the attractive or repulsive nature of the interactions. If we consider that V_p represents the electrostatic potential around a molecule at point r, then for a test charge q, the energy of the electrostatic interaction is $q \times V_p$. It is demonstrated in most physics texts that the relationship between electrostatic potential and charge density is given by Gauss's law in the form known as Poisson's equation:

$$\nabla^2 V_p = -4\pi\rho \tag{9.12}$$

where ∇^2 is the Laplacian defined in Chapter 1 and ρ is the charge density at r. Returning to the definitions in Eq. 9.5, the electrostatic potential can be written as

$$V_p = \sum_{A} \frac{Z_A}{|r - R_A|} - \sum_{ij} P_{ij} \int \frac{\Psi_i \Psi_j}{|r - r'|} dr' \tag{9.13}$$

where Z_A is the charge on nucleus A located at R_A and r' is a dummy integration variable.

Most semiempirical and *ab initio* programs calculate electrostatic potentials (ESPs) as developed by Besler, Merz, and Kollman.[10] Three-dimensional plots displaying isovalues of the surface are available for programs with graphics capabilities. Other programs allow one to plot contour plots of sections taken through the molecule and surrounding volume of space. Once the quantum mechanical ESPs have been calculated, atom-centered monopole charges may be calculated, which may be compared to charges calculated by the Mullikan method, among others. Besler et al.[10] reported that MNDO results with a scaling factor of 1.42 more closely replicated *ab initio* 6-31G* values than did AM1 values (also scaled).

A comparison of Mullikan and ESP atom charges for propyne is instructive (Table 9.2). Electronic interactions in this molecule were first studied at the *ab initio* level by Newton and Lipscomb[11] who concluded that there was no

Table 9.2. A Comparison of Electronic Charges for Propyne Comparing Mullikan and ESP Values by MNDO and RHF/6 – 31G* Methods

H e
C - C≡C-H
H d c b a
H

	MNDO		6 – 31G*	
Atom	Mullikan	ESP	Mullikan	ESP
a	0.156	0.426	0.278	0.355
b	−0.122	−0.546	−0.477	−0.593
c	−0.187	0.028	0.119	0.176
d	0.148	−0.183	−0.521	−0.289
e	0.001	0.090	0.200	0.117

delocalization of the methyl C–H electrons into the π orbitals of the triple bond, i.e., no ground state hyperconjugation. The MNDO ESP values have been scaled by a factor of 1.42, as recommended.[10] What is important here is that the MNDO Mullikan values show C-2 to be more negatively charged than C-1. The MNDO ESP values correspond with the known preference for electrophiles to attack first at C-1 (e.g., the Markownikoff addition of hydrogen chloride). This result is an indicator that Mullikan charges must be used with caution. It is also well known that *ab initio* charges by either method are sensitive to the choice of basis sets. However, the correct conclusion would be drawn from either method in the *ab initio* calculation.

The ESP atom charges for acryloin are shown below. Below that is a representation of an isovalue plot of an ESP surface produced by the program CAChe. One gets a somewhat different point of view, in that the charges on the oxygen–carbon backbone are now submerged within the charges on the hydrogens, which are nearer the surface.

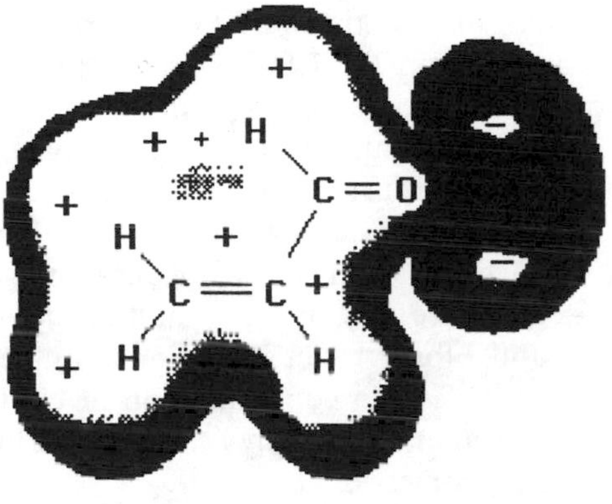

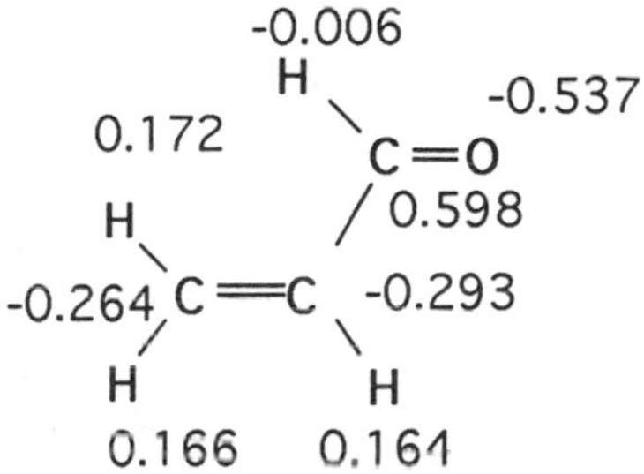

The uses of ESPs in regard to chemical reactivity have been reviewed by Politzer and Murray,[12] who point out that ESP values are particularly useful in predicting electrophilic additions but require further consideration when nucleophilic addition is encountered. Methods for population analysis have also been recently reviewed.[13]

9.6 Open-Shell Systems, Allyl—A Case Study

To this point, we have considered closed-shell electronic systems in which electrons are placed into orbitals in pairs. The restricted Hartree-Fock (RHF) method was built on this premise. If one wishes to examine open-shell systems, such as free radicals or triplet states, a modified approach is required. The earliest of these was to split the spin system into two independent systems of α and β spins. This technique is referred to as the spin-unrestricted Hartree-Fock

(UHF) method. Two sets of MOs may be defined as follows:

$$\Psi_i^\alpha = \sum_{i=1}^{N} c_{\mu i}^\alpha \psi_\mu \qquad \Psi_i^\beta = \sum_{i=1}^{N} c_{\mu i}^\beta \psi_\mu \tag{9.11}$$

and two sets of Fock equations are now required:

$$\sum_{\nu=1}^{N} (F_{\mu i}^\alpha - \varepsilon_i^\alpha S_{\mu\nu}) c_{\mu i}^\alpha = 0$$

and

$$\sum_{\nu=1}^{N} (F_{\mu i}^\beta - \varepsilon_i^\beta S_{\mu\nu}) c_{\mu i}^\beta = 0 \quad \mu = 1, 2, 3, \ldots, N \tag{9.12}$$

The computational time for UHF calculations is longer, since two sets of coefficients and orbitals must be computed. Because the UHF method included twice the number of orbitals and determinants, the variational principle requires that the UHF energy will be lower than the RHF energy for those systems which can be compared. In practice, for closed-shell systems, UHF and RHF energies are frequently in very good agreement. One consequence of the UHF method is that the resultant wave functions are not true eigenfunctions of the total spin operator. As a result, higher spin states may mix into doublet states, leading to spin contamination. UHF calculations include values for the $\langle S2 \rangle$ operator, which is equal to $S(1 + S)$ where S is the spin multiplicity; 0 for a singlet state, 1/2 for a doublet state, and 1 for a triplet state. For a doublet free radical, $\langle S^2 \rangle$ should be $0.5(1 + 0.5) = 0.75$. Spin contamination in UHF calculations leads to higher values for this term and is usually accompanied by a decrease in the accuracy of the computed geometry.

Two methods have been employed in addressing this problem, while retaining the convenience of the RHF method. For the purposes of this discussion, let us consider the five-electron systems shown below:

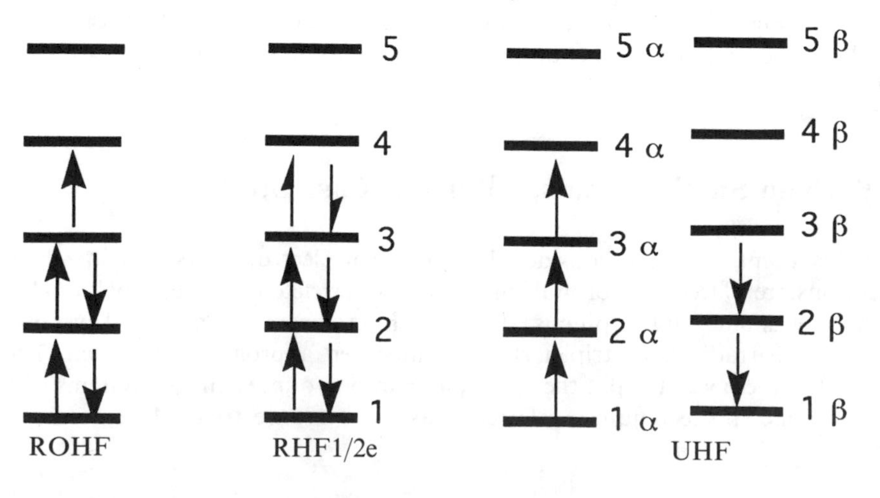

On the right are the two levels associated with the α and β spins in the UHF calculation. In 1968, Dewar and co-workers[14] proposed a scheme in which the lone electron was put into the HOMO orbital of the radical by using a mass equivalent to 1/2 of an electron in each spin configuration. In this fashion, the requirement of all paired electrons was maintained. The Dewar half-electron method has been incorporated in MOPAC. Because of the artifice used, one cannot calculate spin densities by this method.

Concurrent with the above, McWeeny and Diercksen[15] published a different approach to the problem. Again the normally filled orbitals were treated separately from the orbital with the lone electron. Self-consistent perturbation theory was applied to the two density matrices. This method has been dubbed the restricted open-shell Hartree-Fock method (ROHF). It has not been incoporated into MOPAC but rather into the Gaussian family of programs.[16] A great deal will be said about these in the subsequent chapter, but it can be said here that these programs also support semiempirical calculations with the AM1 or PM3 Hamiltonians utilizing the STO-3G basis set. This also will be discussed in the next chapter. For comparison purposes, the allyl free radical has been calculated by all three methods. For the MOPAC UHF calculation, the keywords were PM3 UHF, whereas for the RHF calculation, just PM3 was used. In both cases, the program recognizes the doublet nature of the allyl species when given with no charge designation. The ROHF calculation was carried out with the #ROHF/PM3 designation using Gaussian 92 for Windows. The details of establishing Gaussian job files will be given subsequently. Doublet multiplicity was stated. The UHF value for $\langle S^2 \rangle$ was 0.9394, indicating some spin contamination. The half-electron method is used automatically by MOPAC if UHF is not specified. As can be seen, the RHF and the ROHF methods give heats of formation and C–C bond lengths in close agreement with each other and with the literature value for the heat of formation. As noted above, the UHF value for H_f is lower in value than the RHF values.

Table 9.3. Calculations of the Allyl Free Radical by Three PM3 Methods

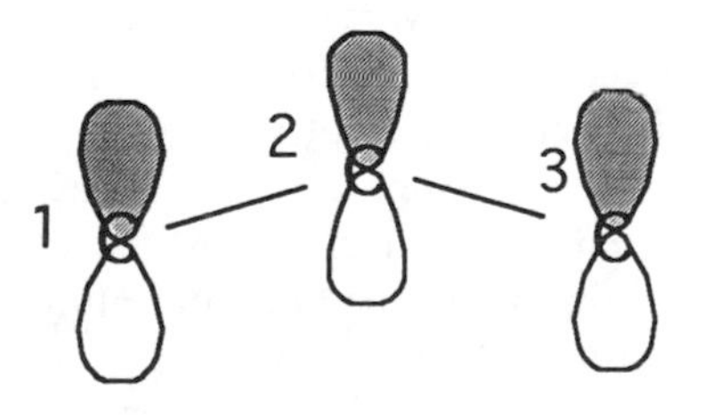

	RHF(1/2e)	ROHF	UHF	Experimental
H_f (kcal/mol)	39.6	39.6	31.2	39.1
$C_1 - C_2(C_2 - C_3)$ Å	1.3733	1.3727	1.3824	

9.7 Normal Coordinates, Force Calculations, and Vibrational Spectra

In considering the problems of multiple conformational minima in Chapter 8 we described a potential energy surface (PES), which varied as bonds were rotated and set into new vibrational patterns. The search methods for various conformational local minima and for a global minimum were briefly described. Let us now carry the argument another step forward.

Given a molecule composed of N atoms, each allowed a vibratory motion, $3N$ degrees of freedom will be required to describe the overall motion of the molecule. Since the molecular translational and rotational motions do not determine the position of the molecule on the PES, we require only $3N - 6$ degrees of freedom ($3N - 5$ for a linear molecule) to describe the internal vibrations of the atoms. The usual procedure is to reference these motions to the center of molecular mass in what is called a *normal coordinate* treatment. In response to a question about the meaning of normal coordinates in a molecule, a physical chemist colleague once provided the following: Imagine a molecule with the atoms held together by springs. Now compress all the springs and atoms together into a compact ball and throw the molecule into the air. The motions described by the various atoms will correspond to the normal coordinates for the molecule. These vibrations are referred to as normal modes. For a many-atom system, it is customary to set the equations of motion in matrix form.

For large molecules, it is usual to assume that the energy associated with the vibration of two atoms making up a given bond is a harmonic oscillator of the Hooke's law type. No forces are operating on the atoms if they are at rest at the equilibrium bond distances. As the bond is stretched, a restoring force proportional to the displacement is generated ($f = -kx$). The first derivative of the energy with respect to the geometry defines the applied force, i.e., $(d\,PE/dx) = kx$. If the equilibrium position is conveniently set at the zero point of the PE scale, then integration gives $PE = \frac{1}{2}kx^2$, a parabola describing the bottom portion of the Morse potential curve for bond stretching. In the higher-energy vibrational levels, the anharmonicity of the potential energy becomes increasingly important. In very exact work, one sometimes sees a correction applied for this effect. In order to obtain the normal vibrations, it is necessary to calculate the second derivative of the molecular energy with respect to each of the geometric parameters. In a matrix format, these second derivatives are called Hessians, each term of which corresponds to a force constant for a given bond. For a molecule at a minimum on the PES, all values of the force matrix are positive. For $3N$ normal coordinates, there will be $3N - 6$ vibrations. Not all of these may have importance, since a requirement is that vibrational lines are only infrared active when there is a change of dipole moment associated with the given vibration. The actual spectrum may have more lines than those computed by the normal coordinate treatment, since various combination and overtone lines may be observed.

Table 9.4. Computation of the Vibrational Frequencies for Acetic Acid

Frequency (cm^{-1})	Transition Dipole	Experimental Values and Intensities[a]
54	0.0181	93-m
397	0.3109	534-m
469	1.1134	581-m
507	0.7270	642-s
571	1.0060	657-s
962	0.1677	847-w
975	0.1975	1084-w
1010	0.3100	1182-s
1240	0.6728	1264-m
1353	0.4572	1382-m
1387	0.1639	1430-*
1387	0.1049	1430-*
1457	2.2343	1382-m
1981	4.6512	1788-vs
3084	0.2303	2996-vw
3091	0.2418	2944-vw
3175	0.2143	3051-vw
3855	0.4600	3583-m

[a]Vw, very weak; w, weak; m, medium; s, strong; vs, very strong
*These bands occur on a shoulder.

The calculated and experimental lines for acetic acid are given in Table 9.4. All 18 line frequencies are calculated, but some of the intensities (proportional to the calculated transition dipole moment) are quite small. The values below were calculated by MOPAC with the PM3 Hamiltonian. The keywords PM3 FORCE were used on line 1. MOPAC analyzes which vibrations are associated with a given frequency. Some commercial programs incorporating MOPAC will display these vibrations in the form of movies.

9.8 Rate Theory and Transition Structures

Although it is very nice to be able to calculate accurately the optimum geometries and the energies of a wide variety of molecules, chemists would like very much to be able to calculate reaction rates and to predict the structures of intermediates and transition states. Perhaps the situation with regard to intermediates should not be put into the same category as transition states, since high-energy intermediates do represent minima on the PES and should be amenable to the same techniques as stable starting materials and products. This is a somewhat conditional statement, in the sense that neither molecular mechanics nor semiempirical methods are easily parameterized to handle such structures. Slow but steady progress on this problem is being made. In part, the problem is made difficult by the lack of spectral and thermodynamic data on reactive intermediates.

According to absolute rate theory, in a one-step reaction, reactants A and B are in equilibrium with a high-energy configuration that represents a saddle point on the free energy surface. This point is called the *transition state*, and the free energy difference between this point and the free energies of the reactants constitutes the activation energy.[17] An unknowable fraction of a material achieving the transition state proceeds in the forward direction to give products, the remainder regressing to starting materials. The mathematical derivation of the absolute rate constant for the reaction requires the assumption of a thermodynamic equilibrium between the ground state reactants and the transition state. Not everyone in science accepts the theory of absolute rates, and one reason is the contradiction in terms offered by this assumption of an equilibrium that must be constantly suffering the perturbation of material leaking off to product(s). The pragmatic approach, however, would point out the great utility provided by absolute rate theory in how chemists, particularly organic chemists, think about reaction pathways and energetics. It has proven worthwhile to swallow certain inconsistencies.

At this point it is necessary to introduce some terminology that is now appearing with increasing frequency. Absolute rate theory discusses the transition state in terms of a maximum in the reaction path on the free energy surface

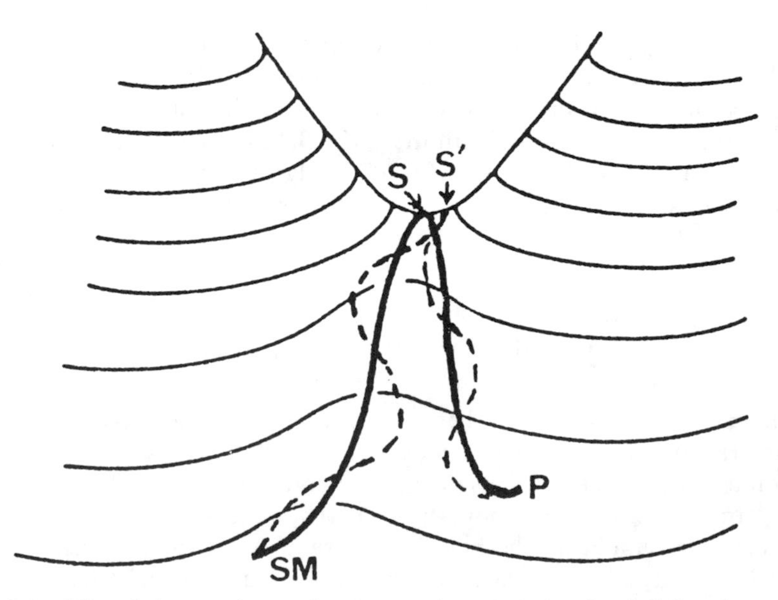

Figure 9.2. Plot of the reaction pathway over the energy barrier defining the activation free energy. The starting point SM, transition state S, and products P are noted. The dotted trajectory represents the path of a "saddle" calculation passing through a transition sturcture S′.

(see Fig. 9.2). The reactants and products appear as minima on that surface. Calculations by the methods now available may give heats of formation (enthalpies) corresponding to the transition state, but these will differ from the free energy by the $-T\Delta S$ term. Depending on the method of computation, corrections may need to be made also for zero point energies and changes in vibrational energies which depend on the temperature of the reaction. This is less true for AMPAC or MOPAC calculations than for *ab initio* calculations, since these programs are parameterized for 25° with the zero point energies included. Potential energy maxima calculated by current techniques are called "transition structures." If the reaction enthalpy is large and if entropy changes in the region of the maximum are small, transition structures may closely resemble true transition states.

Another point to be made regards stationary points on the PES. Products, reactants, and the transition structure all represent stationary points on the PES, which, for this purpose, may be defined as having the first derivative of the energy with respect to all $3N - 6$ independent degrees of freedom equal to zero. However, the second derivatives of the energy for starting materials and products will have positive values (the force constants). For the transition structure, there is one and only one negative root for the force matrix. In the frequency calculation, this negative root leads to a negative (or imaginary) frequency. The degree of freedom having this negative second derivative corresponds to the reaction coordinate. This is the test for a valid transition structure.

The above concept may be rephrased as follows: For stable entities representing local or global minima, the curvature of the PES is everywhere positive. For a transition structure, the PES curvature is everywhere positive except in one direction, and that is along the reaction coordinate.

9.9 Methods of Finding the Transition Structure

There are basically three methods for identifying approximate transition structures. Once located, the validity of the structure must be tested by the method just described. The first method is that of the intelligent guess. Whenever possible, the guess should be based on related, known cases. If the initial guess does not give a single negative root to the force matrix and a single negative frequency, then one must either change the guess and try again or employ some variation of the second method to be described.

The second method is a more systematic approach in which a reaction coordinate (RC) is chosen, and a variable corresponding to a change along the RC is varied in a systematic fashion until an energy maximum is found. This method may be called "path following." In a variation of a path search, two variables may be selected, allowing an exploration of a portion of the PES in two dimensions. This extension is called a "grid search." These methods may be used to augment the intelligent guess method.

9.10 The Sn2 Displacement Reaction—A Case Study[18]

The MOPAC file below exemplifies the attack of a chloride ion of methyl fluoride in an Sn2 reaction. Note that no Hamiltonian is specified, so the program defaults to an MNDO calculation. The reaction equation is shown first. The chloride ion is placed 20 Å from the carbon to start. The −1 tells the

$$Cl^- + CH_3F \rightleftharpoons CH_3Cl + F^-$$

```
1   CHARGE = -1
2   SN2 REACTION
3
4    C     0.00    0      0.00    0      0.00    0    0    0    0
5    F     1.40    1      0.00    0      0.00    0    1    0    0
6    H     1.10    1    109.50    1      0.00    0    1    2    0
7    H     1.10    1    109.50    1    120.0     1    1    2    3
8    H     1.10    1    109.50    1   -120.0     1    1    2    3
9    Cl   20.0    -1    127.3     1    180.0     1    1    2    3
10   0     0.00    0      0.00    0      0.00    0    0    0    0
11  10.0   5.0     4.0    3.0     2.9    2.8     2.6  2.5
12   2.4   2.3     2.2    2.1     2.0    1.9     1.8  1.7  1.6
```

program this is a path calculation, and this distance is to be varied according to the list after the row of zeros. The latter tells the program where the Z-matrix stops. The Cl–C distance is sent through a set of fixed distances the balance of the molecule is minimized, and the new geometry and energy are stored in the output file. The pertinent portion follows:

C–Cl	C–F	HF	C–Cl	C–F	HF
20.0	1.35	115.9	2.3	1.40	−95.47
10.0	1.35	−116.6	2.2	1.41	−87.9
5.0	1.35	−119.1	2.1	1.42	−78.7
4.0	1.36	−120.4	2.0	1.45	−67.6
3.0	1.36	−118.4	1.9	1.47	−54.2
2.9	1.37	−117.1	1.8	1.52	−37.4
2.8	1.37	−115.5	1.7	3.25	−42.3
2.6	1.37	−110.3	1.6	3.40	−27.3
2.5	1.38	−106.5			
2.4	1.39	−101.6			

The above data contain some surprises, which will be discussed a little later. However, note that as the Cl–C bond begins to approach a normal bonding length, the energy, which had been decreasing, starts to oscillate. This may be taken as indicating that we are in the region of the transition structure. If we reverse this calculation, setting the C–F distance to 20Å, and watch the energy and C–Cl distance change as the fluoride approaches the carbon, we will see approximately the same behavior. From these data we can estimate the not unreasonable values of C–Cl = 1.95 Å and C–F = 2.20 Å as models for the Sn2 transition structure.

MOPAC offers several options for reversing the normal minimization process, i.e., the structure climbs the energy hill to a maximum. In this case, Baker's eigenvalue-following routine was chosen.[19] The MOPAC file below has the guessed carbon–halogen bond lengths and the keyword TS, which specifies the eigenvalue-following method is to be applied. Note that the −1 has been changed to 1 so that no path calculation is signaled. The output structure from this calculation is given below. The heat of formation was

```
1     CHARGE=-1 TS
2     SN2 REACTION Model of TS for chloride displacing fluorine
3
4  C   0.00000000 0       0.0000000  0        0.0000000  0  0  0  0
5  F   2.20000000 1       0.0000000  0        0.0000000  0  1  0  0
6  H   1.09488577 1      81.4973574  1        0.0000000  0  1  2  0
7  H   1.09462701 1      80.7093613  1      119.4786568  1  1  2  3
8  H   1.09483780 1      80.4718614  1     -119.8379020  1  1  2  3
9  Cl  1.95242823 1     179.3375343  1     -167.8686194  1  1  2  3
```

−43.06 kcal/mol. This structure was then input to a force calculation by replacing the keyword TS by FORCE. The force calculation gave one negative root to the force matrix and one negative frequency, thereby meeting the criteria for a transition structure.

It is instructive to compare this result with a high-level *ab initio* (MP2/6 − 31 + + G**) calculation for the same reaction, as is done in Figure 9.3.[19] Unfortunately, *ab initio* energies cannot be compared directly to heats of formation, though methods to accomplish this transformation are now appearing in the literature.

```
1     CHARGE=-1 TS
2     SN2 REACTION
3
4  C   0.00000000 0       0.0000000  0        0.0000000  0  0  0  0
5  F   2.16430821 1       0.0000000  0        0.0000000  0  1  0  0
6  H   1.09455656 1      81.6996038  1        0.0000000  0  1  2  0
7  H   1.09455017 1      81.7066730  1      119.9793574  1  1  2  3
8  H   1.09453922 1      81.7004887  1     -120.0267748  1  1  2  3
9  Cl  1.96772963 1     179.9886294  1     -167.8486868  1  1  2  3
10
```

The above reaction has been carried out at 25° in the gas phase. When taken together with the path calculations, one can draw a reaction coordinate plot for the gas phase Sn2 reaction consistent with current mechanistic thinking. Ion energies in the gas phase are very high, as no solvation can take place. As

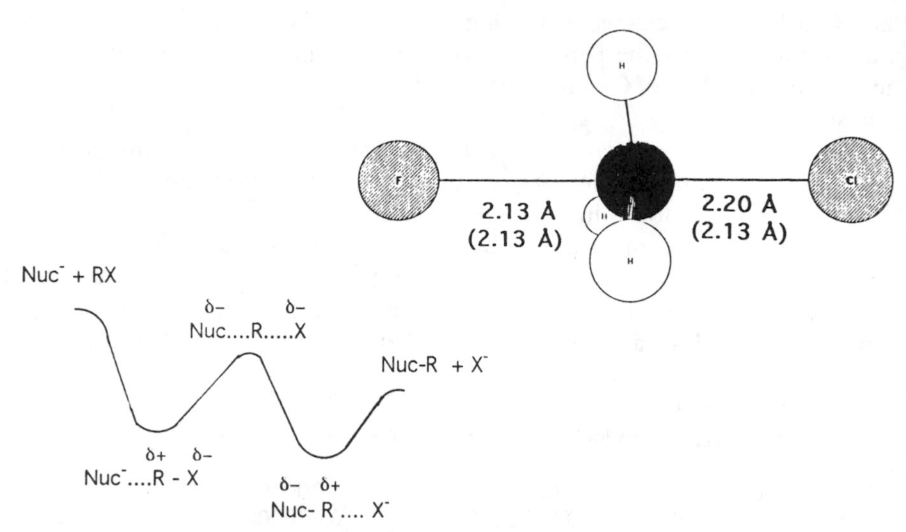

Figure 9.3. Ball-and-stick model of the MNDO Sn2 transition structure and reaction coordinate plot. (MNDO and *ab initio* bond lengths are compared.)

the ion approaches the neutral substrate, ion–dipole interactions lower the energy. As the reaction proceeds, a transition structure is passed and the separation of the newly formed ion–dipolar species begins.

The third method is called a "saddle" calculation.[20] In this method, the structures of the starting materials and the products are each given in the MOPAC format. The two sets must be separated by a row of zeros, as shown below, and the end of the life is also so designated (see below). The calculations are best carried out in Cartesian coordinates, so the keyword XYZ is added. The keyword SADDLE specifies the mode of calculation, which is by the

CHARGE = −1 XYZ SADDLE
SN2 REACTION

C	0.00	0	0.00	0	0.00	0	0	0	0	
F	1.40	1	0.00	0	0.00	0	1	0	0	
H	1.10	1	109.50	1	0.00	0	1	2	0	
H	1.10	1	109.50	1	120.0	1	1	2	3	
H	1.10	1	109.50		−120.0	1	1	2	3	
Cl	5.0	1	127.3	1	180.0	1	1	2	3	
O	0.00	0	0.00	0	0.00	0	0	0	0	
C	0.00	0	0.00	0	0.00	0	0	0	0	
F	5.0	1	0.00	0	0.00	0	1	0	0	
H	1.10	1	109.50	1	0.00	0	1	2	0	
H	1.10	1	109.50	1	120.0	1	1	2	3	
H	1.10	1	109.50	1	−120.0	1	1	2	3	
Cl	1.80	1	127.3		180.0	1	1	2	3	
O	0.00	0	0.00	0	0.00	0	0	0	0	

eigenvector-following technique. One very necessary (and sometimes quite frustrating) condition is the requirement of carrying the starting structure into the final structure with a one-to-one correspondence in atom numbers. The energies and geometries of the starting and final structures are first determined, and a vector of length $3N$ is defined by the difference in geometries of these two structures. The scalar value of this vector is the BAR. The BAR is systematically reduced as new minimizations are carried out. A constraint maintaining the distance of the new structure from the other geometries is applied. Depending on whether the new structure is higher in energy than the previous structures, the process cycles until a structure of maximum energy is reached. The output structure should then be reminimized with the keyword TS, and a force calculation carried out as before. The application of this method to our Sn2 case above is illustrated by the input file.

The refined transition structure agrees very well with our path exercise. The heat of formation found was −42.96 kcal/mol with a C–F distance of 2.15 Å and a C–Cl length of 1.98 Å.

9.11 Caveat

The transition structure located by a saddle calculation needs to be examined carefully to see if it meets a reasonable (intuitive) expectation for a transition structure for the reaction being studied. Sometimes rather strange structures result from this method. One should also realize that transition structures with flexible or rotatable groups like methyl may not always meet the criteria of a single negative root and a single negative frequency. Often simply rotating the methyl to reduce a steric interaction will produce the desired result. The MOPAC manual discusses more complex ways to address this problem.

Each of the MOPAC studies to this point has referred to molecules at 25° in a vacuum. Although this is a perfectly reasonable way to treat neutral nonpolar molecules or even (in most cases) free radical processes, ionic reactions are more usually carried out in solvents with appreciable polarity. In principle, one could produce a solvent model by simply adding a large collection of solvent molecules around the solute of interest. As you can imagine, the computational problems associated with this approach would rapidly overwhelm any computer capability. Just as there has been a continuous search for new ways to find stationary maxima, there has been a concerted effort to build a correction for solvent polarity into the calculations. The most useful of these are based on work by Nobel Laureate Lars Onsager during the early decades of this century. Onsager developed the "reaction field model," in which a polar molecule is placed within a spherical cavity surrounded by the medium, which is assigned a uniform dielectric constant. The molecular dipole moment induces a dipole moment in the solvent, which, in turn, interacts to modify the energy of the solute. For polar molecules in a polar solvent, the result is a net stabilization.

Recently, Klamt and Schürmann[21] have described a new version of the Onsager approach (COSMOS), which has been incorporated into the latest version of MOPAC (MOPAC93). The solute molecule is placed inside a cavity in the form of a regular polygon. The medium outside the polygon is assigned a dielectric constant ε. The COSMOS program is triggered by the keyword ESP, which is assigned the dielectric constant value of the solvent in question. Keywords are available to alter the size of the cavity and to determine the number of sides for the polygon. This approach has been demonstrated to be quite effective for solvents of high dielectric constant such as water. The amino acid glycine is normally calculated to be stabler in the free amino acid form in the absence of solvent. However, when the calculation is carried out with COSMOS set at the dielectric constant of water, the zwitterion form is the stabler.

Application of COSMOS to the saddle calculation above can be carried by altering the first line to read: CHARGE = −1 XYZ SADDLE NSPA = 60 GRADIENTS EPS = 78.4. The NSPA keyword specifies the number of surfaces to be used in constructing the polygon. The initial and final distances for the halide ions were 5 Å from the alkyl carbon. The energies (kcal/mol) are summarized below. As is very evident, the presence of the solvent greatly alters

	First Ion–Dipole	Transition Structure	Second Ion–Dipole	ΔHF	$\Delta HF^\dagger$
No solvent	−119.13	−47.93	−43.00	−76.1	71.2
Water	−218.51	−167.50	−170.10	−48.4	51.0
Differences	99.4	111.6	127.1		

the energy picture. Fluoride is a very small ion (a hard nucleophile) and therefore is solvated strongly by water, compared to the larger, more polarizable chloride ion. Water makes the overall reaction less exothermic, but by lowering the activation energy significantly, enhances the rate of displacement.

References

1. (a) J. A. Pople, D. P. Santry, and G. A. Segal, *J. Chem. Phys.*, **43**, S129 (1965); (b) J. A. Pople and G. A. Segal, *ibid.*, **43**, S136 (1965); (c) J. A. Pople and D. L. Beveridge, *Approximate Molecular Orbital Theory*, McGraw-Hill, New York, 1970.

2. See Chapter 8, ref. 11.

3. Dr. A. J. Holder, Semichem, 12716 West 66th Terrace, Shawnee, KS 66216. See also Chapter 7, ref. 7.

4. For MOPAC see Chapter 7, ref. 7; Chapter 7, ref. 8; and Chapter 8, ref. 2; MOPAC93 © Fujitsu Limited.

5. M. J. S. Dewar, E. G. Zoebisch, E. F. Healy, and J. J. P. Stewart, *J. Am. Chem. Soc.*, **107**, 3902 (1985).
6. J. J. P. Stewart, *J. Comp. Chem.*, **10**, 210 (1989).
7. (a) Cambridge Scientific Computing, Cambridge, MA.
8. Orb Draw from Serena Software, Bloomington, IN.
9. W. J. Hehre, L. Radom, P. v. R. Schleyer, and J. A. Pople, *Ab Initio Molecular Orbital Theory*, John Wiley & Sons, New York, 1986.
10. B. H. Besler, K. M. Merz, Jr., and P. A. Kollman, *J. Comp. Chem.*, **4**, 431 (1990).
11. M. D. Newton and W. N. Lipscomb, *J. Am. Chem. Soc.*, **89**, 4261 (1967).
12. P. Politzer and J. S. Murray in *Reviews in Computational Chemistry*, K. B. Lipkowitz and D. B. Boyd, eds., Vol. 2, VCH Publishers, New York, 1991.
13. S. M. Bachrach, *ibid.*, Vol. 5, 1994.
14. M. J. S. Dewar, J. A. Hashmall, and C. G. Venier, *J. Am. Chem. Soc.*, **90**, 1953 (1968).
15. R. McWeeny and G. Diercksen, *J. Chem. Phys.*, **49**, 4852 (1968).
16. See Chapter 12, ref. 7.
17. S. Glasstone, K. J. Laidler, and H. M. Eyring, *The Theory of Rate Processes*, McGraw-Hill, New York, 1941.
18. Z. Shi and R. J. Boyd, *J. Am. Chem. Soc.*, **111**, 1575 (1989).
19. J. Baker, *J. Comp. Chem.*, **7**, 385 (1986).
20. M. J. S. Dewar, E. F. Healy, and J. J. P. Stewart, *J. Chem. Soc., Faraday Trans. II*, **3**, 227 (1984).
21. A. Klamt and G. Schürmann, *J. Chem. Soc. Perkin Trans 2*, 799 (1993).

CHAPTER

10

Molecular Modeling—*Ab Initio* and Density Functional Methods

As computer power has grown, so has the use of *ab initio* methods of computation for organic structures and mechanisms.[1] One would like to say politely at this point that many articles appearing in the current literature imply that such calculations carry a weight of authority not to be found in the semiempirical methods. But as one gains experience with the *ab initio* methods, it quickly becomes evident that they too are filled with assumptions and levels of accuracy. In fact, published examples now exist in which one set of *ab initio* results contradicts others carried out under different assumptions.[2]

The most immediate contrast with semiempirical methods is the fact that all integrals are evaluated *de novo*, i.e., from scratch. This means there is no parameterization of the integrals by experimental data or by intelligent guessing. Just as with our earlier methods, it is assumed that electronic wave functions are unaffected by nuclear motions (Born-Oppenheimer approximation). The evaluation of these integrals greatly extends the computational work required to accomplish the desired goal. As a consequence, early *ab initio* work was largely confined to small molecules. Although larger systems are studied today, computational times mount rapidly with the number of heavy atoms, and one is often confined to choosing less than the desired computational level to arrive at the goal within a rational time frame.

The computational cycle in an *ab initio* calculation is very similar to that employed in the semiempirical methods, in that an initially guessed structure is used as the starting point, and the first calculation is done at a low level such as an extended Hückel calculation. At this point, two considerations come into play. The first is the choice of basis sets to be used to represent the MOs in the calculation. One must also consider what will be an adequate level of theory, SCF, or is correlation required, and for open-shell systems whether a restricted or unrestricted method is best. For neutral hydrocarbon molecules, the RHF Hamiltonian is often quite adequate, at least as a starting point. Once these choices have been specified, the program begins the iterative cycle of calculating energies and geometries until convergence is reached, at which point the data are printed out.

10.1 Basis Set Choices

Early approaches to *ab initio* calculations attempted to use Slater-type orbitals (STOs) of the form presented in Chapter 9. It was quickly found that the evaluation of the resulting one- and two-centered integrals presented a major computational problem. Subsequently, a number of workers started to explore the replacement of the STO by a mathematically more tractable combination of Gaussian-type orbitals (GTOs).[3] A typical three-dimensional Gaussian has the form

$$x^l y^m z^n e^{-ar^2}$$

where r is the vector distance between the fixed point A and a variable position characterized by the Cartesian coordinates x, y, and z. The exponents l, m, and n vary with the orbital levels, and a is a positive number determining the orbital size. In contrast to the STO, the squared term in the GTO eases the computational difficulties referred to above. Early practice was to replace the STO with a linear combination of Gaussian functions. Ultimately, it was shown that viable results usually could be obtained with three Gaussian terms. These STO-3G orbitals were used in many early *ab initio* studies and are still used today for rapid survey calculations before spending time on higher levels of calculation. It should be pointed out as a rule of thumb that computational time varies approximately as the fourth power of the number of basis functions. STO-3G is referred to as a minimum basis set, because it uses only the number of orbitals required to accommodate the number of electrons for the atom in question. Lithium and beryllium, for instance, require only 1*s* and 2*s* orbitals. However, in order to maintain the spherical symmetry of the atoms boron through neon, the full set of 1*s*, 2*s*, $2p_x$, $2p_y$, and $2p_z$ must be used. As with semiempirical MOs, there are scale factors to be introduced in determining useful orbitals. In some studies this was carried out by minimizing the energy of each atom, but a more usual approach is to refine the scale factors using the energies and geometries of small molecules. Thus, the claim of no arbitrary parameters in *ab initio* calculations is seen to be somewhat of a stretch in terminology.

While matters of terminology are being considered, two other definitions should be mentioned. Single Gaussian functions, called *primitives*, are commonly used in fixed linear combinations, termed contracted Gaussians, and basis sets with more primitives give more accurate orbital descriptions, but at a computational cost. Another term that has appeared with increasing frequency follows from the definition of a molecular orbital, i.e., a volume of space occupied by one or two electrons. "Unoccupied orbitals" is an oxymoron, though the term LUMO for lowest unoccupied orbital continues in use. The term *virtual orbitals* appears increasingly in the literature.

Minimal basis sets suffer because of their inability to adjust to charges and lone electron pairs that may be part of the structure. Compare water to the hydronium ion or ammonia to the ammonium ion; at the STO-3G level, the same set of orbitals are used for the neutral and charged species. Reasonably, the orbitals of the positively charged ions should be contracted when compared to their neutral counterparts. To borrow some terminology, STO-3G orbitals are too "hard," too nonpolarizable.

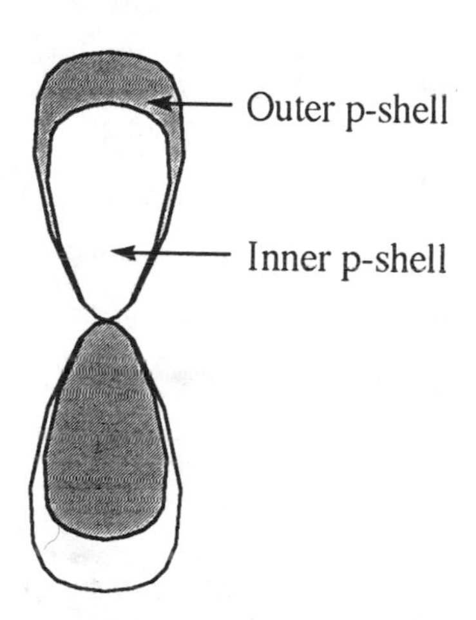

One approach to a solution is to increase the number of primitive Gaussians used to describe the orbital and in essence split the orbital into an inner "hard" core and an outer "soft" or more polarizable shell. If the inner-shell orbitals are represented by a single orbital made up of a set number of Gaussians, while the valence-shell orbitals are split into two parts each, the basis set is said to be a split-basis set. The phases here are shown reversed for pictorial purposes only. For a second row element, the orbital array is as follows:

$1s$

$2s', 2p'_x, 2p'_y, 2p'_z$

$2s'', 2p''_x, 2p''_y, 2p''_z$

The separation between the inner and outer functions can be adjusted by parameter choice. Typical split-valence basis sets are the 3-21G and 6-31G sets.[1] The first number gives the number of primitive Gaussians in the inner-core orbitals. The split-valence orbitals are given with the number of Gaussians for the inner portion (2 or 3 in these cases) and the number for the outer portion (1 in each case). For elements lithium through neon, the configuration for a split-basis set is as shown above. In contrast, even the inner-shell orbitals may also be split, in which case the basis set is described as a *double zeta* set. The reader encountering these concepts for the first time may suddenly realize that he has wandered rather far afield from the descriptions of orbitals given in introductory organic chemistry textbooks.

The justification for the departure from the elementary picture of atomic and molecular orbitals is that the structures and energies that result are considerably more accurate that those provided by the more naive theories.

Two additional basis-set corrections should be mentioned before leaving the topic. Both split and double zeta sets lead to orbitals centered about the nucleus. Molecules with small or highly strained rings as well as those which are very polar require orbitals that allow a nonuniform distribution of charge. This can be effected by adding additional *d* functions or a combination of *d* and *p* functions. The effect of the addition of a higher *d* function to a *p* function might be represented as follows:

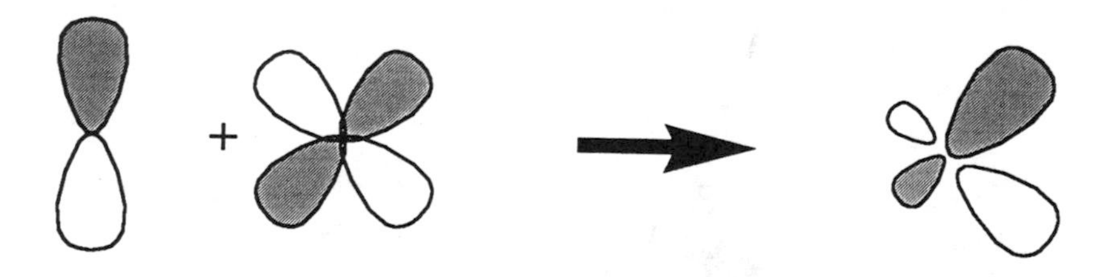

As we have observed before, the combination of two functions produces two new orbitals. Not shown is the second, mirror-image orbital to the above. One way of designating these polarized basis sets is by listing the functions added, i.e., 6-31G(*d*, *p*), which adds d functions to the nonhydrogen atoms and p functions to the hydrogen atoms. This notation is sometimes given as 6-31G**. Since these are approximations, the model is expected to improve as the basis becomes more flexible.

In order to cope with atoms where the electron density is far removed from the nucleus, it is necessary to add yet more *s* and *p* functions of large extent to the basis set. If the additional terms apply only to heavy atoms and not to hydrogen atoms, then a single + sign is used, i.e., 3-21 + G and 6-31 + G. If *s* functions are added to the hydrogens as well, then a + + is designated, 3-21 + + G or 6-31 + + G. The effect on geometries and energies is given for the water molecule in Table 10.1. As can be seen, expansion of the basis set carries a penalty in computational time.

Table 10.1. MO Calculations on Water by Various Methods

Method and Basis Set	O–H Å	H–O–H degrees	Energy	Basis Orbitals	Primitives	CPU Secs[b]
Mopac AM1 STO	0.9609	103.67	–59.24[a]	6		1.3
HF/AM1 STO-3G	0.9612	103.50	–0.09443[a]	7	21	91
HF/STO-3G	0.9893	99.97	–74.9659	7	21	375
HF/3-21G	0.9668	107.66	–75.5859	13	21	326
HF/6-31G	0.9496	111.54	–75.5859	13	30	443
HF/6-31G*	0.9474	105.52	–76,0107	19	36	442
HF/6-31 + G*	0.9475	106.50	–76.0177	23	40	481
MP2/6-31G	0.9747	109.26	–76.1132	13	30	449
MP2/6-31G*	0.9686	103.99	–76.1968	19	36	617
MP2/6-31 + G*	0.9708	105.50	–76.1132	23	40	681
MP4/6-31G*	0.9702	103.96	–76.2073	19	36	2660
Exp Values	0.957	104.5	–57.8[a]			

[a]All values are for full optimizations. The MOPAC and experimental energies for water are heats of formation in kcal/mol. The AM1 STO-3G result is calculated by Gaussian 92 for Windows[7] which gives all energies in hartrees. This value in kcal/mol is –59.26. All other energies are total energies and cannot be compared directly to the first two entries.

[b]Mopac and Gaussian 92 for Windows calculations were carried out on a 486-DX operating at 66 MHz.

10.2 Computational Methods

Virtually all *ab initio* calculations start out at the Hartree-Fock SCF level either restricted (RHF) for closed-shell systems or unrestricted (UHF) for open-shell species. For many purposes, such as geometry optimization of neutral molecules, this is the method of choice. But as mentioned in Chapter 9, the Hartree-Fock method ignores electron correlation effects, and these can become important in studying charged species, highly strained molecules, transition states, and calculations of spectral frequencies. Two methods of correcting this will be mentioned here. Since both of these methods start with a Hartree-Fock calculation and then a series of corrective calculations is appended, they are known as post-SCF methods.

Corrections for configuration interaction (CI) were mentioned briefly in Chapter 9, and such calculations are applied as post-SCF calculations in *ab initio* work. The most common CI correction starts with Hartree-Fock SCF orbitals and then adds to the ground state wave functions those involving single and double substitutions(excitations) from the ground state wave functions. Higher-order CI corrections involve an increasing number of virtual and filled orbitals with a corresponding increase in computational time.

The second approach to the electron correlation problem is to add perturbation terms to the Hatree-Fock Hamiltonian. The most used of the several methods available is the Møller-Plesset perturbation theory, which may be truncated at several orders of perturbation. The effect on the water structure of applying this theory is given in Table 10.1, where it can be seen that there is a significant lowering of the energy at the 6-31G* level when passing from RHF to RMP2 and RMP4 Hamiltonians. The effects of correlation become particularly important when one is involved in comparing points on a PES such as the comparisons between molecular ground states and related transition states.

10.3 Practical Matters

Ab initio calculations give total energies in atomic units or hartrees. The hartree is defined as the energy of Coulombic repulsion between two electrons separated by a distance of one bohr radius ($a_0 = 0.529$ Å), i.e.,

$$1 \text{ hartree} = \frac{e^2}{a_0}$$

One hartree equals 627.51 kcal. The total energy includes electron repulsion, nuclear repulsion, and electron–nuclear attraction. Semiempirical programs are calibrated to give total energies in electron volts but also present the heat of formation, which can be compared directly with experimental measurements. In contrast, the total energies calculated by *ab initio* methods are computed at absolute zero (0° K). In order to arrive at the room-temperature energy, it is necessary to carry out a frequency calculation and use the partition functions found there to approximate the value at room temperature. Furthermore, these values need to be corrected for zero-point energy vibrations, a value which is also obtained from the frequency calculation. These matters are handled internally as part of the parameterization in semiempirical calculations. Recently, methods of converting *ab initio* total energies to heats of formation have been a subject of much investigation.[4] A simple method of correlating *ab initio* energies for hydrocarbons with experimental heats of formation has been devised by W. C. Herndon.[5] Two relations are given here:

For (6-31G*) calculations:

$$\Delta HF = 590.577436 * E_{(6\text{-}31G*)} + 22373.755(N_C) + 337.107(N_H) \qquad (10.1)$$

For (6-31G*//STO-3G):

$$\Delta HF = 589.754489 * E_{(6\text{-}31G*//STO\text{-}3G)} + 22340.834(N_C) + 336.590(N_H) \qquad (10.2)$$

The designation (6-31G*//STO-3G) is a type of notation often found in papers with *ab initio* results. It is frequently expeditious to carry out the geometry minimization at a lower level of basis set and then do a single-point calculation of the energy based on that geometry using a higher-level basis set. In this case an STO-3G RHF geometry was used for each hydrocarbon, followed by a 6-31G* single-point energy minimization. These equations were parameterized with 65 examples drawn from alkanes, alkenes, alkynes, their cyclic counterparts, and a series of aromatic hydrocarbons as well as alkyl-substituted hydrocarbons. Strained molecules were included in the above set.

10.4 Available Programs

The writing of computer programs for both semiempirical and *ab initio* calculations is a daunting task. Many early versions of these programs were made available through the Quantum Chemistry Program Exchange (QCEP) at the University of Indiana[6] and are still quite serviceable. Newer versions in many instances are now available only through commercial sources.[8] A widely used set of *ab initio* programs has been produced from the laboratories of J. A. Pople at Carnegie-Mellon University under the general name of Gaussian plus a version number. Versions through Gaussian 85 are still available through QCEP. Gaussian 92 and Gaussian 92 for Windows[9] are available through Gaussian, Inc. Since the latter is the program with which the author is familiar, all further examples will be based on Gaussian 92 for Windows, though most of the remarks also apply to Gaussian 92 and earlier versions in this series.

Gaussian input files rather resemble those for MOPAC. As a typical example, let us use the case of the water molecule. The numbers along the left-hand side are for reference purposes only and would not be included in the file input. Line 1 gives a name to what is termed the "checkpoint" file. This is a binary file that holds molecular data that can be passed to extended calculations as frequency calculations or correlation corrections, or to restart some calculations.

Line 2 starts with a # character and is termed "the route." Here keywords are selected that control the type of calculation performed. This example requests a restricted geometry Hartree-Fock optimization using the 3-21G basis. Line 4 is an informational title. Both the route and the title can be multiple lines and are terminated by a blank line (lines 3 and 5 in this case). There is a wide range of keywords that can be used on this line, depending on the type of calculation wanted. The users should consult the Gaussian 92 manual for other keywords and options. Line 6 gives the charge on the molecule (zero in the case above), and the second digit gives the spin multiplicity. All lines below this point are part of the Gaussian Z-matrix. The format is very much like the MOPAC format, except that in this example, the distances and angles are given in the form of variables, which are then assigned values below. Gaussian programs will also read atom positions in Cartesian coordinates. However, it is rather difficult to find programs that will accept graphical input of structures and save the files as Cartesian coordinates. The

```
        Comments
 1  Checkpoint file   %chk=formic
 2  Route              #RHF/3-21G opt
 3  blank
 4  description       formic acid optimization at the 3-21G level
 5  blank
 6                    O  1
 7                    C  1   r21
                      O  2   r32     1   a321
                      H  2   r42     3   a423    1   d4231   0
                      H  3   r53     2   a532    1   d5321   0
 8                    Variables:
 9                    r21          1.23
                      r32          1.36
                      r42          1.10
                      r53          0.97
                      a321       117.58
                      a423       112.35
                      a532       110.58
                      d4231     -179.98
                      d5321        0.13
10  blank
```

author has found it convenient to generate his structures first in PCMODEL (Chapter 8) and then minimize them with MOPAC including the keyword "AIGOUT," which will cause the program to generate a Gaussian-type Z-matrix as part of the output of the archive file (name.arc). This file will have to have the first several lines edited, which can be readily accomplished using the Windows file editor. There are other programs reputed to support the generation of Z-matrix files in the correct format, but the author is not familiar with specific examples. The final line of the input file must end with a blank line.

This type of calculation is frequently employed when a prior structural minimization has taken place at a lower level of computation, the initial structure having been saved in the checkpoint file. If we assume the geometry obtained above to be satisfactory, the calculation at, say, the 6-31G** level could be accomplished as follows:

```
1       %chk=formic
2        #RHF/6-31G(d,p)  Geom=checkpoint
        test
3
4       formic acid  (6-31G**//3-21G)
5
6       0 1
7       blank line
```

On a 486-DX (66 MHz) computer, the 3-21G optimization required 11 min 9 s and gave an energy of −187.700199 hartrees, and the 6-31G** single-point calculation required 3 min 36 s with an energy of −188.768145 hartrees. In contrast, a direct 6-31G** optimization required one hour, giving an energy minimum of −188.770566 hartrees. Thus, the (6-31G**//3-21G) computation saved 45 minutes of computational time compared to the full 6-31G** optimization, with a net energy difference of 1.5 kcal/mol in favor of the longer calculation. Minor differences in geometry also were found. So it is a case of "you pay your money, and you take your choice." For other examples the reader is directed to reference 1. The two sections of the (6-31G**//3-21G) calculation could have been linked to carry the whole process out as one calculation. All that is needed is to put a line following the blank line for the first portion as follows:

```
--link 1--
```

This line is followed immediately by instructions for the single-point calculation.

10.5 Potential Energy Surfaces

As with semiempirical calculations, *ab initio* programs offer a variety of methods for searching potential energy surfaces. The Gaussian programs have incorporated the linear synchronous transit method of Halgren and Lipscomb.[10] This method preceded in time the saddle calculation method described in Chapter 9 but is sufficiently similar to not require further description. In both cases, the input data are the geometries of reactants and products. In both, a pathway is followed that leads to a geometry of maximum energy somewhere near the saddle point on the reaction coordinate. This point may not be exactly at the correct position on the col between the valleys. Refinement methods (keyword OPT=TS) are available, and the nature of the discovered transition structure is ascertained by a frequency (FREQ) calculation, which must give one and only one negative eigenvalue to the force constant matrix as before. The LST input file for the 1,2-migration of a hydrogen atom in the ethyl free radical is given here:

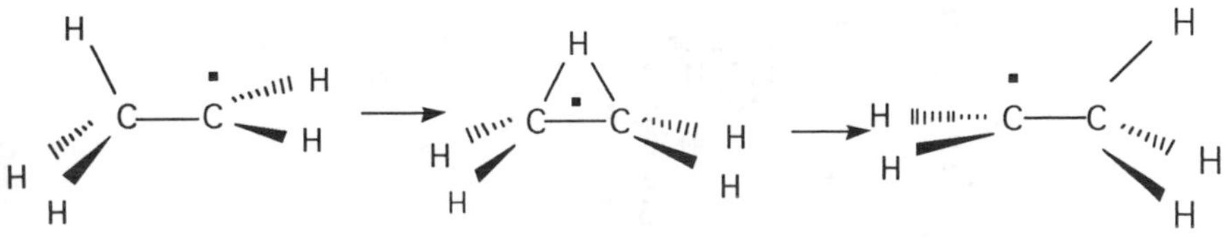

```
#UMP2/6-31g(d)  LST  GUESS=AL-
WAYS

ETFR1 rearrangement to ETFR2

0 2
C
C    1  r21
H    2  r32        1  a321
H    2  r42        1  a421       3  d4213
H    2  r52        1  a521       3  d5213
H    1  r61        2  a612       3  d6123
H    1  r71        2  a712       3  d7123

VARIABLES:
r21            1.5158
r32            1.0897
r42            1.0865
r52            1.0865
r61            1.0829
r71            1.0829
a321         111.258
a421         110.839
a521         110.8337
a612         118.8965
a712         118.9009
d4213        119.8702
d5213       -119.8481
d6123        -90.0000
d7123         90.0000

ETFR2

0 2
C
C    1  r21
H    2  r32        1  a321
H    2  r42        1  a421       3  d4213
H    2  r52        1  a521       3  d5213
H    1  r61        2  a612       3  d6123
H    1  r71        2  a712       3  d7123

VARIABLES:

r21            1.5162
r32            2.1646
r42            1.0828
r52            1.0828
r61            1.0867
r71            1.0866
a321          27.9853
a421         118.9028
a521         118.938
a612         110.8179
a712         110.8545
d4213         90.0000
d5213        -90.0000
d6123       -119.8486
d7123        119.8689
```

The keyword GUESS indicates how the initial Hartree-Fock wave function will be established, and a wide variety of options are available. In this case, GUESS = ALWAYS requests that a new initial guess be generated at each step in the optimization. The default condition utilizes the result of the preceding SCF calculation as a guess for the next point.

The structure located by the LST calculation above is the symmetrically bridged structure shown. Ignoring zero point energy corrections for the moment, the difference in energy between the classical ethyl radical and the bridged entity is 65.5 kcal/mol. Refinement was not carried out in this case, as the system had been studied by *ab initio* methods some years earlier.[11] It was shown there that the zero point energy correction lowers the energy barrier by 5 to 7 kcal/mol, depending on the exact level of the calculation. Values of 46 to 53 kcal/mol were given, again depending on the computational level. In comparison, a semiempirical UHF/AM1 saddle calculation gave a symmetrically bridged transition structure also, which, upon refinement, was 42 kcal/mol above the classical starting radical. It is clear from the above results that hydrogen migration in alkyl radicals such as ethyl cannot be an important process. In fact, the loss of a hydrogen atom to form ethylene has a lower energy barrier.[10]

The picture changes when the migrating group is phenyl. Semiempirical saddle calculations and *ab initio* LST calculations agree that the rearrangement of the 2-phenylethyl radical passes through a structure corresponding to an unsymmetrically bridged 2-phenylethyl radical, forming a symmetrically bridged cyclohexadienyl species as shown below.

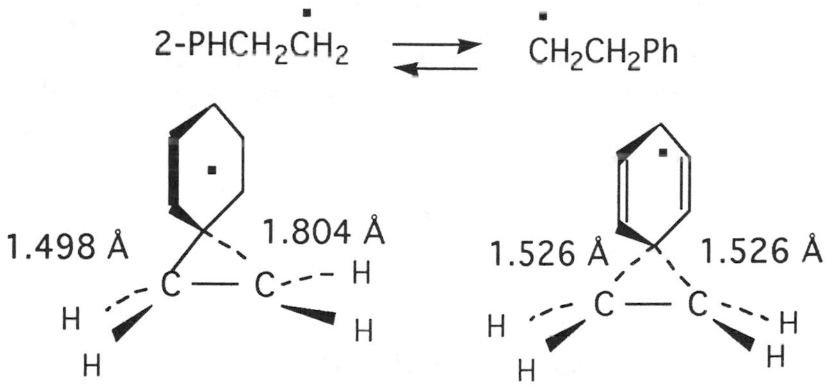

$$2\text{-PHCH}_2\dot{\text{C}}\text{H}_2 \rightleftharpoons \dot{\text{C}}\text{H}_2\text{CH}_2\text{Ph}$$

1.498 Å 1.804 Å 1.526 Å 1.526 Å

If one can arrive at a reasonable guess at the transition structure, the OPT = (TS, CALCFC) keywords offer a possible identification of the correct transition structure. The CALCFC requests that all force constants be calculated rather than estimated. Another possibility is to carry out a grid search over the area of potential surface near the guessed structure (keyword SCAN). Either one or two of the molecular geometry variables may be selected for variation in the scan. The name of the variable, the initial value, the numbers of points to be scanned, and the increment size are specified.

Once a suitable transition structure has been identified, it may be desirable to show that the structure does lie on the path from reactants to products. Calculation via the Intrinsic Reaction Coordinate is supported by both semiempirical and *ab initio* programs. In either case, the calculation starts with the refinement of the transition structure and a force calculation to confirm the validity of the structure. Starting from the transition structure, the calculation carries through a series of structures back to the reactants or forward to the products. The program chooses a normal coordinate from the force calculation and follows the potential energy change toward the reaction initiation (or terminus) point. In this fashion, a continuous sequence of structures defining the entire reaction pathway may be generated. Force calculations carried out at each terminus allow the calculation of the activation energies for the overall process. In *ab initio* calculations, the zero point energy correction must be applied.

10.6 A Case Study—Ethylene

Some of the utility of *ab initio* calculations for determining molecular geometries and energies has been indicated. Methods appropriate to exploring molecular PESs have also been reviewed. There are other useful bits of information which can be derived from these calculations. Some of these will be indicated here, with specific application to the ethylene molecule.

The first ionization potential for an organic molecule may be defined as the energy required to remove the electron from the HOMO to infinity. One method might be to calculate the energy for the neutral species and then for the cation radical formed by the removal of one electron. Although this method gets one "into the ball park," it should be kept in mind that any inaccuracies in determining the two respective HOMO energies will be additive in the calculated ionization potential. The HF/6-31G* energies for ethylene and the ethylene cation radical are as follows:

$$\dot{C}H_2 - CH_2^+ \quad -77.58078 \text{ hartrees}$$

$$CH_2 = CH_2 \quad \underline{-78.37438}$$

$$\Delta E = \quad 0.450937 \text{ hartrees}$$

$$\underline{\times\ 27.21} \text{ eV/hartree}$$

$$12.27 \text{ eV ionization potential}$$

Ionization potentials are measured experimentally by photoelectron spectroscopy. The molecule under investigation is irradiated with photons generated either by a gaseous discharge or from soft x-rays, and the kinetic energy of the ejected electron is measured. The ionization potential is the difference in energy between the irradiation source and the kinetic energy. The experimental first ionization potential for ethylene is 10.51 eV.

A longstanding theorem of Koopman maintains that one may approximate the ionization potential of an electron from its eigenvalue. This approximation has been shown to be quite accurate for the first two or three ionization potentials. The HOMO eigenvalue for ethylene by the 6-31G* calculation is at −0.3743 hartrees. Converting this value by the conversion factor above gives 10.18 eV, a value closer to the correct value than that given above.

It is known that the lowest energy absorption in the electronic spectrum results from the transition of an electron from the HOMO to the LUMO of the molecule. The LUMO value for ethylene in our calculation above is 0.18391 hartrees. The HOMO-LUMO gap then is 0.558 hartrees. The wavelength corresponding to this energy can be shown to be

$$\lambda_0 = \frac{c}{\nu_0} = \frac{hc}{\Delta E} = \frac{28625}{\Delta E\,\text{kcal}}$$

At 627.51 kcal/hartree, the HOMO-LUMO gap corresponds to 350 kcal, and the computed transition should fall at 82 nm. In fact, the lowest energy absorption band falls at 170 nm. This difference is understandable, as experience has shown that virtual eigenvalues are routinely on the high side. A comparison of the above results with semiempirical AM1 calculations gives 1055 eV by Koopman's theorem and a calculated absorption at 104 nm.

Literature values for the heat of formation of ethylene are 12.5 and 14.5 kcal/mol. The AM1 value is 16.5 kcal/mol. The HF/6-31G* value can be calculated from Eq. 10.1 as 12.2 kcal/mol.

Textbooks of physical chemistry or spectroscopy make the point that molecular vibrations are quantized. The various levels of vibrational energy have an envelope resembling the well-known Morse curve. To a good approximation, the lower levels can be treated as a quantized harmonic oscillator model for which the envelope is a parabola. As higher energy levels come into play, there is an increasing amount of anharmonicity due to the true envelope shape. Both *ab initio* and semiempirical methods can be employed to calculate vibrational spectra (infrared and Raman) by obtaining the force matrix and applying the harmonic oscillator approximation. The force matrix is composed of the second differentials of the energy with respect to all x, y, and z components of motions for each of the atoms in the molecule. Diagonalization of this matrix gives the force constants from which the harmonic frequencies can then be calculated. The keyword for this calculation is FORCE in MOPAC and FREQ in Gaussian 92. The number of normal vibrations is readily calculated. Motions for the $3n$ atoms of ethylene equal 18. Since there

Table 10.2. Infrared Frequencies for Ethylene

Exp (cm^{-1})	AM1	MP2/6-31G*	Becke3LYP/6-31G* (a)
926	834	849	835
949	1068	989	973
1444	1412	1520	1496
2990	3185	3213	3152
3106	3218	3322	3247

[a]A result from DFT to be discussed below.

are three translational and three rotational degrees for the molecule as a whole, which do not affect the vibrational modes, there remain 12 normal vibrational modes ($3n - 6$). For a linear molecule, the number of modes is $3n - 5$. In order for a vibration to produce an infrared absorption, there must be an accompanying change in the dipole moment. As a result, not all calculated normal vibrational frequencies will produce a spectral line. Furthermore, a variety of combination and overtone bands may exist, leading to more than the calculated number of fundamental lines. We will not go into the computation of these frequencies.

Table 12.2 gives the calculated frequencies for ethylene by several different methods. Only those vibrational lines with a finite intensity are listed, though the calculations give frequencies for all 12 normal vibrations. The nature of the parameterization makes the semiempirical results above appear equal to or better than those of the more complex calculations. The calculated *ab initio* results at the Hartree-Fock level are too large by about 11%, a correction that is often put into effect as an empirical adjustment. This results, in part, from deficiency in the inclusion of correlation terms as well as the failure to correct for the anharmonicity. These effects are roughly equal in magnitude. The correlation deficiency can be corrected by doing a CI correction or by perturbation corrections such as MP2 or MP4. The MP2 method has been used here with the 6-31G* basis set, as this represents about the minimum level useful for computational purposes. Calculation by a density functional method is also given. This subject is taken up in the following section. Density functional computation improves on the electron correlation as an inherent part of the calculation.The nature of the parameterization makes the semiempirical results above appear equal to or better than those of the more complex calculations.

10.7 Density Functional Theory

Density functional theory (DFT) has been around for some time.[12] However, its application to problems of interest to chemists has only recently begun to attract attention.[13] Basically, the origins are to be found in early attempts to treat electron densities in atomic systems as though the density had gaslike propeties. The mathematical approach is related to the calculus of variations and includes an understanding of statistical thermodynamics. Many of the early successes of DFT may be attributed to physicists working on problems with solids. It is only in recent years that applications to organic chemistry have begun to appear, and the jury is out on the ultimate utility of this method (or methods) for organic structure and mechanisms. These questions will be answered in the next few years, as a commercially available program associated with the program Gaussian 92 is now available for use.[8]

The first question many organic chemists will ask is, what is a functional? We know that a function is a mathematical prescription for going from a variable, say x, to a number $f(x)$. Similarly, a functional is a prescription for going from a function to a number, i.e., $F[f(x)]$. A case in point might be the Schrödinger equation $\int \Psi H \Psi = E$ where the terms to the left of the equal sign constitute a functional that operates on the wave function to give the number associated with E.

In DFT, at least as practiced by the Gaussian programs, the exchange term in the Hartree-Fock equation is replaced by an integral (a functional) which is a function of the electron density. The practical effect is to introduce a two-part functional that handles both the exchange and correlation terms. Thus, the immediate and most obvious advantage of DFT is the inclusion of electron correlation, which is ignored in conventional Hartree-Fock calculations. Although this alteration imposes an additional computational step, these integrals are computationally more easily handled. Because of the time requirement for computational setup, DFT methods applied to small molecules require more time than comparable Hartree-Fock SCF methods. However, for large molecules the setup time is a smaller fraction of the total time for computation, and accurate results may be obtained with a material saving in real time. A variety of exchange and correlation functionals are now available in the literature. Comparisons of Hartree-Fock and DFT results are beginning to appear.[13,14] An example comparing *ab initio* and DFT methods is given in Table 10.3. The density functional used in this study was Becke's three-parameter exchange functional[15a,b] and the correlation functional of Lee, Yang, and Parr.[15c] This choice seems to offer results comparable with the HF/6-31G* and MP2/6-31G* methods. Methanol (CH_3OH) has $(3 \times 6) - 6 = 12$ normal modes and frequencies. The lack of symmetry produces an infrared spectrum in which all 12 lines are to be found. The experimental values and calculated results by the *ab initio* and DFT methods indicated are given in Table 10.3. The average error for the MP2/6-31G* results was 108 cm^{-1}, whereas that for the DFT result was 70 cm^{-1}, a considerable improvement. No corrections were applied to these calculations, though it is customary to scale *ab initio* values to allow for the imperfect corrections for electron correlation. The computer time for the above calculations was about twice as long for the DFT calculation. This is due to the proportionally greater length of the setup time for the DFT calculation. This factor becomes of less importance when working with large molecules, as noted above.

Table 10.3. Experimental and Calculated Frequencies for Methanol by the MP2/6-31G* and Becke3LYP/6-31G* Methods[a]

	Exp (cm^{-1})	MP2	Becke3LYP
	3681.00	3795.00	3736.00
	3000.00	3222.00	3168.00
	2960.00	3143.00	3081.00
	2844.00	3076.00	3032.00
	1477.00	1578.00	1537.00
	1477.00	1564.00	1519.00
	1455.00	1539.00	1498.00
	1345.00	1418.00	1396.00
	1165.00	1203.00	1171.00
	1060.00	1112.00	1085.00
	1033.00	1082.00	1052.00
	295.00	351.00	361.00
C–O	1.421	1.424	1.419
O–H	0.963	0.970	0.968
C–O–H	108.0	107.4	107.6

[a]Experimental geometry taken from ref. 1 with bond lengths in Å and angles in degrees.

As a final example, a comparison of the geometries and singlet and triplet energy gap for methylene as calculated by DFT and *ab initio* methods will be given. It has been known for a number of years that the carbene, methylene, can be formed in either of two electronic states.

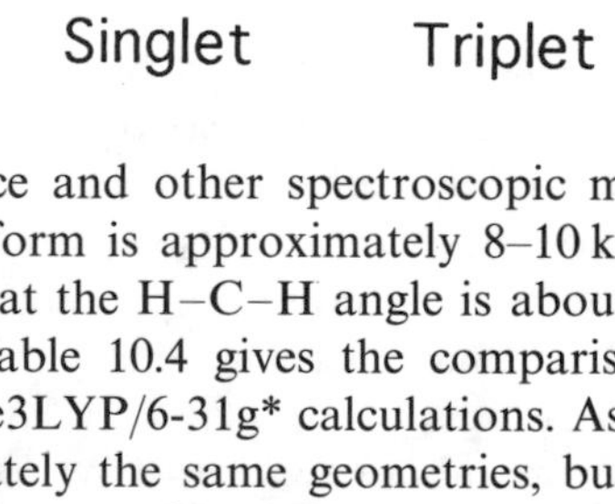

Electron spin resonance and other spectroscopic means have been used to show that the singlet form is approximately 8–10 kcal/mol higher in energy than the triplet, and that the H–C–H angle is about 103° for the singlet and 136° for the triplet. Table 10.4 gives the comparison data for unrestricted MP2/6-31g* and Becke3LYP/6-31g* calculations. As can be seen, both calculations give approximately the same geometries, but the DFT calculation is somewhat closer to the experimental energy range. Again, because of the differences in setup time, the DFT calculation required about twice the time of the MP2 calculation.

Table 10.4. Calculations on Singlet and Triplet Methylene

	UMP2/6-31G*		Becke3LYP/6-31G*	
	S	T	S	T
C–H Å	1.109	1.078	1.119	1.082
H–C–H deg	101.95	131.54	100.28	133.11
$\Delta E(S - T)$ kcal mol	—	17.7	—	13.7

References

1. For a detailed account, see W. J. Hehre, L. Random, P. v. R. Schleyer, and J. A. Pople, *Ab Initio Molecular Orbital Theory*, John Wiley & Sons, New York, 1986.

2. M. W. Wong, A. Pross, and L. Radom, *Israel J. Chem.*, **33**, 415 (1993).

3. For an account of the properties of Gaussian functions, see I. Shavitt, *Methods in Computational Physics*, Vol. 2, John Wiley & Sons, New York, 1963.

4. See N. L. Allinger, L. R. Schmitz, I. Motoc, C. Bender, and J. K. Labanowski, *J. Am. Chem. Soc.* **114**, 2880 (1992) and references therein.

5. W. C. Herndon, University of Texas-El Paso, *Chem. Phys. Lett.*, **234**, 82 (1995).

6. Quantum Chemistry Program Exchange (QCPE), Creative Arts Building, Indiana University, Bloomington, IN 47405.

7. CaCHe, see Chapter 7, ref. 8, and SPARTAN from Wavefunction, Inc., 18401 Von Karman Avenue, Suite 370, Irvine, CA 92715.

8. M. J. Frisch, G. W. Trucks, M. Head-Gordon, P. M. W. Gill, M. W. Wong, J. B. Foresman, J. B. Johnson, H. B. Schlegel, M. A. Robb, E. S. Replogle, R. Gomperts, J. L. Andres, K. Raghavaachari, J. S. Binkley, C. Gonzalez, R. L. Martin, D. J. Fox, D. J. DeFrees, J. Baker, J. J. P. Stewart, and J. A. Pople, Gaussian, Inc., Carnegie Office Park, Building 6, Pittsburgh, PA 15106.

9. "Windows" is a registered trademark of the Microsoft Corporation, just as "Gaussian 92" is for Gaussian, Inc.

10. T. A. Halgren and W. N. Lipscomb, *Chem. Phys. Lett.*, **49**, 225 (1977).

11. L. B. Harding, *J. Am. Chem. Soc.* **103**, 7469 (1981).

12. (a) *The Single-Particle Density in Physics and Chemistry*, N. H. March and B. M. Deb, eds., Academic Press, New York, 1987; (b). *Density-Functional Theory of Atoms and Molecules*, R. G. Parr and W. Yang, eds., Oxford University Press, New York, 1989.

13. *Density Functional Methods in Chemistry*, J. K. Labanowski and J. W. Andzelm, eds., Springer-Verlag, New York, 1991.

14. B. G. Johnson, P. M. W. Gill, and J. A. Pople, *J. Chem. Phys.* **98**, 5612 (1992).

15. (a) A. D. Becke, *J. Chem. Phys.*, **98**, 5648 (1993); (b) A. D. Becke, *Phys. Rev.*, **A38**, 3098 (1988); (c) C. Lee, W. Yang, and R. G. Parr, *Phys. Rev.*, **B37**, 785 (1988).

APPENDIX

Matrices and Determinants

As we pass through the early years of our education, we become used to handling numbers as single entities in such common operations as addition, subtraction, and multiplication. It is only when one attempts courses dealing with linear algebra or physics that one becomes aware of a whole expanded scheme of mathematics that functions with collections or arrays of numbers and sometimes abstract symbols.

A matrix is an array of numbers or symbols such as those which follow:

$$\begin{pmatrix} a_{11}b_{12}c_{13}d_{14} \\ a_{21}b_{22}c_{23}d_{24} \\ a_{31}b_{32}c_{33}d_{34} \\ a_{41}b_{42}c_{43}d_{44} \end{pmatrix} \qquad (a_1b_2c_3d_4) \qquad \begin{pmatrix} a_1 \\ b_2 \\ c_3 \\ d_4 \end{pmatrix}$$

The left-hand matrix is described as a 4×4 square matrix, and the other two matrices are described as a row vector and a column vector, respectively. We will examine some of the properties of matrices after first discussing some of the properties of determinants, which can be derived from such matrices. Through both discussions we will work only with square arrays, as these are the only ones of importance to the materials we have considered in the text.

Consider the set of simultaneous equations below:

$$a_1x + b_1y = k_1$$
$$a_2x + b_2y = k_2$$

By cross multiplying each equation and subtracting, it is possible to arrive at the values for x and y as follows:

$$x = \frac{k_1 b_2 - k_2 b_1}{a_1 b_2 - a_2 b_1} \quad \text{and} \quad y = \frac{a_1 k_2 - a_2 k_1}{a_1 b_2 - a_2 b_1}$$

The expression $a_1 b_2 - a_2 b_1$ is frequently cast in the form of the determinant, i.e.,

$$a_1 b_2 - a_2 b_1 = \begin{vmatrix} a_1 a_2 \\ b_1 b_2 \end{vmatrix} \tag{A.1}$$

The use of the vertical straight lines indicates a determinant as opposed to a matrix. Determinants are always square and may be associated with a corresponding square matrix. However, determinants can be reduced to a number and matrices cannot. Although certain operations are common to both matrices and their determinants, others are not.

The value of the determinant may be obtained as in Eq. A.1 if it is of the second order. Larger determinants can be reduced by the method of expansion of minors or by the use of appropriate computer programs.

As an example of expansion, refer to the secular determinant for butadiene in Chapter 2, page 17. The order of the original 4×4 determinant is first reduced to a series of 3×3 determinants, which, in turn, are then reduced to 2×2 determinants, which are expanded as above. These smaller determinants are called minors of the original. The minor of the element a_{12} is formed by suppressing all the elements in the first row and in the second column, reducing the original determinant or matrix by an order of 1. Note the change in sign of the expanded terms in an alternate fashion when the suppressed element is at the head of columns 2,4,..., etc. It should also be pointed out:

$$\begin{vmatrix} x & 1 & 0 & 0 \\ 1 & x & 1 & 0 \\ 0 & 1 & x & 1 \\ 0 & 0 & 1 & x \end{vmatrix} = x \begin{vmatrix} x & 1 & 0 \\ 1 & x & 1 \\ 0 & 1 & x \end{vmatrix} - \begin{vmatrix} 1 & 1 & 0 \\ 0 & x & 1 \\ 0 & 1 & x \end{vmatrix} = 0$$

$$x^2 \begin{vmatrix} x & 1 \\ 1 & x \end{vmatrix} - x \begin{vmatrix} 1 & 1 \\ 0 & x \end{vmatrix} - \begin{vmatrix} x & 1 \\ 1 & x \end{vmatrix} + \begin{vmatrix} 0 & 1 \\ 0 & x \end{vmatrix} = 0$$

$$x^2(x^2 - 1) - x(x - 0) - (x^2 - 1) = 0$$

$$x^4 - 3x^2 + 1 = 0$$

multiplying a determinant by an element equal to zero makes that whole term equal to zero. Thus, the fourth-order determinant above expands to two third-order determinants. Obviously, this will not be true if the first row contains no zero elements.

Faced with this fourth-order polynomial, several devices may be tried to arrive at the four roots. If the equation can be reduced to a quadratic, the quadratic formula may be applied. Because of the nature of the bonding of carbon in π systems, all roots must lie between ± 3. One can try 0 and all integers readily enough. Frequently, the equation will factor. Finally, one may assign values to x and plot the curve to determine the roots. It is helpful to remember that the number of sign alterations between successive coefficients equals the number of negative roots. In our equation there are two such alterations. Consequently, there must be two negative roots.

The equation can be cast, in this case, in the form of a quadratic by setting $y = x^2$, hence the quartic above becomes $y^2 - 3y + 1 = 0$. Applying the formula for the solution of a quadratic equation:

$$y = \frac{-b \pm \sqrt{b^2 - 4ac}}{2a}$$

gives

$$y = \frac{3 \pm \sqrt{9 - 4}}{2}$$

$$y = 0.38, 2.62 \quad \text{and}$$

$$x = \pm 0.62 \text{ and } \pm 1.62$$

As given in Chapter 2, these are the eigenvalues for butadiene.

There are a number of computer programs that will both reduce the secular determinant to its polynomial form and then determine the roots (eigenvalues) of the polynomial. Commonly available spreadsheet programs will often provide a number of capabilities in handling determinants and matrices. The author is familiar with the Mathcad[1] program and provides an example showing the application to the problem above.

Roots to the Secular Equations via Mathcad

Setup the 4×4 secular determinant under the math window choosing matrix (4×4).

$$\begin{bmatrix} x & 1 & 0 & 0 \\ 1 & x & 1 & 0 \\ 0 & 1 & x & 1 \\ 0 & 0 & 1 & x \end{bmatrix}$$

The solution resulting in the fourth order polynomial is carried out in the symbolic portion of the program. Load the SYMBOLIC PROCESSOR and click on one parenthesis placing a blue box around the determinant, go to the SYMBOLIC menu and click on Determinant of Matrix

$x^4 - 3 \cdot x^2 + 1$

$x^4 - 3.0 \cdot x^2 + 1.0 = 0$

$$\begin{bmatrix} 1.6180339887498948482 \\ -.6180339887498948482 \\ .6180339887498948482 \\ -1.6180339887498948482 \end{bmatrix}$$

Retype the polynomial expressing the decimal points as shown since this converts the calculation to a floating point calculation. The equal sign shown is the "symbolic equal sign" which is generated on a Macintosh by depressing the option-command and = keys. Windows operation will require consulting the manual. Click on this expression and place the vertical bar adjacent to one of the x s. Go to the SYMBOLIC menu and select SOLVE FOR VARIABLE. The roots for the polynomial will be given to a far greater number of significant figures than that required for subsequent calculations.

Calculations of Coefficients

Additional information about the butadiene molecule depends on a knowledge of the complete set of MO wave functions. The coefficients may be determined as before, i.e., by substituting each root, one at a time, into the secular equations and solving simultaneously. However, a convenient method is the so-called method of cofactors illustrated below.

In this case, it is possible to arrive at normalized coefficients using the relationships

$$N = \sqrt{\sum \left(\frac{c_n}{c_1}\right)^2} \qquad n \text{ is the column}$$

$$\frac{c_n}{c_1} = + \frac{(\text{cofactor})_n}{(\text{cofactor})_1} \qquad n \text{ is odd}$$

$$\frac{c_n}{c_1} = - \frac{(\text{cofactor})_n}{(\text{cofactor})_1} \qquad n \text{ is even}$$

The ratio of cofactors (minors) for each coefficient must be set up first.

$$\frac{c_1}{c_1} = \frac{\begin{vmatrix} x & 1 & 0 \\ 1 & x & 1 \\ 0 & 1 & x \end{vmatrix}}{\begin{vmatrix} x & 1 & 0 \\ 1 & x & 1 \\ 0 & 1 & x \end{vmatrix}} = 1$$

$$\frac{c_2}{c_1} = \frac{\begin{vmatrix} 1 & 1 & 0 \\ 0 & x & 1 \\ 0 & 1 & x \end{vmatrix}}{\begin{vmatrix} x & 1 & 0 \\ 1 & x & 1 \\ 0 & 1 & x \end{vmatrix}} = \frac{1 - x^2}{x^3 - 2x}$$

$$\frac{C_3}{C_1} = \frac{\begin{vmatrix} 1 & x & 0 \\ 0 & 1 & 1 \\ 0 & 0 & x \end{vmatrix}}{\begin{vmatrix} x & 1 & 0 \\ 1 & x & 1 \\ 0 & 1 & x \end{vmatrix}} = \frac{1}{x^2 - 2}$$

$$\frac{C_4}{C_1} = \frac{\begin{vmatrix} 1 & x & 1 \\ 0 & 1 & 1 \\ 0 & 0 & x \end{vmatrix}}{\begin{vmatrix} x & 1 & 0 \\ 1 & x & 1 \\ 0 & 1 & x \end{vmatrix}} = \frac{-1}{x^3 - 2x}$$

For the root $x = -1.62$

n	C_n/C_1	$(C_n/C_1)^2$	C_n
1	1.00	1.00	0.372
2	1.62	2.62	0.602
3	1.62	2.62	0.602
4	1.00	1.00	0.372

$$\sum \left(\frac{c_n}{c_1}\right)^2 = 7.24$$

$$\sqrt{\sum \left(\frac{c_n}{c_1}\right)^2} = 2.69$$

Thus, the lowest-energy wave function for butadiene is

$$\Psi_1 = 0.372\psi_1 + 0.602\psi_2 + 0.602\psi_3 0.372\psi_4$$

In a similar fashion, the other roots of x can be shown to lead to

$$\Psi_2 = 0.602\psi_1 + 0.372\psi_2 - 0.372\psi_3 + 0.602\psi_4$$
$$\Psi_3 = 0.602\psi_1 - 0.372\psi_2 - 0.372\psi_3 + 0.602\psi_4$$
$$\Psi_4 = 0.372\psi_1 - 0.602\psi_2 + 0.602\psi_3 - 0.372\psi_4$$

The Application of Matrices

Conceptually, one might reduce a secular determinant to a series of minors which would then be evaluated as demonstrated for butadiene utilizing the eigenvalues to determine the MO coefficients. However, a more facile approach is the utilization of a secular matrix with its attendant matrix manipulations. The secular determinants that we have worked with here and earlier in the text can be converted to such matrices by exchanging the x terms on the diagonal by zero. Thus the secular matrix for butadiene becomes

$$\begin{pmatrix} 0 & 1 & 0 & 0 \\ 1 & 0 & 1 & 0 \\ 0 & 1 & 0 & 1 \\ 0 & 0 & 1 & 0 \end{pmatrix}$$

Note the use of the curved vertical lines specifying a matrix. Technically, matrices such as this fall into a more general class known as *Hermitian* matrices. For our purposes, Hermitian matrices will be considered to be real, square, symmetrical arrays. As with determinants, matrices may be specified by a series of running subscripts

$$\begin{pmatrix} a_{11}a_{12}a_{13}\cdots a_{1j} \\ a_{21}a_{22}a_{23}\cdots a_{2j} \\ \vdots \\ a_{i1}a_{i2}a_{i3}\cdots a_{ij} \end{pmatrix}$$

which may be summarized as $\mathbf{A} = (a_{ij})$.

The transpose of a matrix is defined as a new matrix in which the element a_{ij} is interchanged with its corresponding element a_{ji}. Thus if $\mathbf{A}$ is the matrix, then $\mathbf{A}^{\mathrm{T}}$ is its transpose.

$$\mathbf{A} = \begin{pmatrix} a_{11}a_{12}a_{13} \\ a_{21}a_{22}a_{23} \\ a_{31}a_{32}a_{33} \end{pmatrix} \qquad \mathbf{A}^{\mathrm{T}} = \begin{pmatrix} a_{11}a_{21}a_{31} \\ a_{12}a_{22}a_{32} \\ a_{13}a_{23}a_{33} \end{pmatrix}$$

Since the Hermitian matrices we will deal with are real and symmetrical, $\mathbf{A} = \mathbf{A}^{\mathrm{T}}$.

Square matrices can be multiplied by sequentially multiplying rows of the first matrix times the appropriate columns of the second, i.e.,

$$(a_{mm}) * (b_{mm}) = (c_{mm}) \quad \text{and} \quad c_{de} = \sum_{j=1}^{n} a_{dj}b_{je}$$

as an example

$$\begin{pmatrix} a_{11}a_{12} \\ a_{21}a_{22} \end{pmatrix} * \begin{pmatrix} b_{11}b_{12} \\ b_{21}b_{22} \end{pmatrix} = \begin{pmatrix} (a_{11}b_{11} + a_{12}b_{21})(a_{11}b_{12} + a_{12}b_{22}) \\ (a_{21}b_{11} + a_{22}b_{21})(a_{21}b_{12} + a_{22}b_{22}) \end{pmatrix}$$

The operation corresponding to division is not defined for matrices. However, our matrices **A** are all characterized as having an inverse $\mathbf{A}^{-1}$ such that $\mathbf{A} \times \mathbf{A}^{-1} = \mathbf{I}$ where **I** is a square matrix with the same dimensions as **A** with all a_{ii} equal to 1 and all a_{ij} zeros. Mathcad and many spreadsheets will produce inverses for a given matrix.

Finally, to address the problem of deriving eigenvalues and eigenfunctions from the eigenmatrix, we must recognize the existence of matrix similarity transformations. Two matrices **A** and **B** are said to be related by a similarity transformation if a matrix exists such that

$$\mathbf{SAS}^{-1} = \mathbf{B} \tag{A.2}$$

It will be stated here without proof that if **A** is an eigenmatrix and if the columns of **S** consist of the eigenvectors of **A**, then **B** will be the related matrix in the diagonal form with the diagonal elements equal to the eigenvalues of **A**. There are methods of diagonalizing small Hermitian matrices by hand, but it is far more convenient to do so with any of a number of computer programs. Once the eigenvalues are known, Eq. A.2 may be used to determine **S** with its eigenvectors. Staying with our example of butadiene, the application of Eq. A.2 to determine the eigenvalues and then the eigenfunctions by the program MATHCAD is given below.

Mathcad Matrix Diagonalization for Eigenvalues and Eigenvectors

$$\mathrm{M} := \begin{bmatrix} 0 & 1 & 0 & 0 \\ 1 & 0 & 1 & 0 \\ 0 & 1 & 0 & 1 \\ 0 & 0 & 1 & 0 \end{bmatrix}$$

Mathcad requires the secular matrix to be setup from the secular determinant by removing the x s on the diagonal. Note the use of the colon-equals sign which serves to define the matrix M.

$$\mathrm{eigenvals(M)} = \begin{bmatrix} -1.618 \\ 1.618 \\ -0.618 \\ 0.618 \end{bmatrix}$$

The eigenvals(M) = command causes the eigenvalues of the matrix to be determined. Note that the eigenvalues are not printed in ascending order.

$$\mathrm{eigenvec(M, 1.618)} = \begin{bmatrix} 0.372 \\ 0.602 \\ 0.602 \\ 0.372 \end{bmatrix} \qquad \mathrm{eigenvec(M, 0.618)} = \begin{bmatrix} -0.602 \\ -0.372 \\ 0.372 \\ 0.602 \end{bmatrix}$$

$$\text{eigenvec}(M, -0.618) = \begin{bmatrix} 0.602 \\ -0.372 \\ -0.372 \\ 0.602 \end{bmatrix} \qquad \text{eigenvec}(M, -1.618) = \begin{bmatrix} -0.372 \\ 0.602 \\ -0.602 \\ 0.372 \end{bmatrix}$$

To accomplish the same ends, one may choose to use the matrix diagonalization program written in QuickBASIC[2]. This should run on any machine with a BASIC capability. QuickBASIC on Macintosh computers runs best with a 24-bit processor (earlier Macs) or those in which the 24-bit mode can be set. Presumably Apple IIs will run this program, though the author has not tested this premise. The initial code line is required on machine running in MS DOS[2]. The author runs this program on a 486-DX machine using the PRINT SCREEN key to print the output. There is no print statement in the Macintosh version.

```
DECLARE SUB Eigen (EgnVect #(), EgnVal #(), n%)
DEFDBL a-z
DEFINT i-n
INPUT "order of matrix",n
DIM a1(n,n),a2(n,n)
FOR i=1 TO n:FOR j=1 TO n:INPUT a1(i,j):NEXT:NEXT

Eigen a1(),a2(),n
FOR i=1 TO n
 a1(i,1)=a1(i,i)
NEXT

FOR i=1 TO n:PRINT a1(i,1):NEXT
PRINT "Eigenvalues"

FOR i=1 TO n:FOR j=1 TO n:PRINT a2(i,j),:NEXT:PRINT:NEXT
LCOPY
END

'**********************************************************************************************

'Eigen is based on a FORTRAN routine published in Davis (1978).
'It was specifically designed to find the eigenvectors and eigenvalues of a real
symmetrical matrix. 'EgnVect() on input should contain the nxn matrix. On exit it
will contain the 'eigenvectors in descending order. 'On exit EgnVal() contains
the eigenvalues stored columnwise, where column '1 corresponds to the first
eigenvector, etc.

'**********************************************************************************************

SUB Eigen(EgnVect(),EgnVal(),n) STATIC
```

```
  ANORM=0 #
   'Form the identify matrix in EgnVal() & compute threshold.
  FOR i=1 TO n
    FOR j=1 TO n
     IF i=j THEN EgnVal(i,j)=1 #ELSE
EgnVal(i,j)=0 #:ANORM=ANORM+EgnVect(i,j)*EgnVect(i,j)
    NEXT
  NEXT
  ANORM=SQR(ANORM)
  FORM=ANORM*.000000001 #/n
  thr=ANORM
23 thr=thr/n
3  DidRotate%=0     'Set DidRotate flag to false
 'Scan down columns for off-diagonal elements greater than or equal to
the threshold value
  FOR i=2 TO n
    i1=i-1
    FOR j=1 TO i1
      IF ABS(EgnVect(j,i))>=thr THEN     'compute sine and cosine for
rotation
        DidRotate%=-1     'We are rotating so set rotate flag to true
       AL=-EgnVect(j,i)
        AM=(EgnVect(j,j)-EgnVect(i,i))/2 #
       AO=AL/SQR(AL*AL+AM*AM)
       IF AM<0 THEN AO=-AO
        SINX=AO/SQR(2 #*(1 #+SQR(1 #-AO*AO)))
        SINX2=SINX*SINX
        COSX=SQR(1 #-SINX2)
        COSX2=COSX*COSX
        For k=1 TO n                'rotate columns i and j
          If k=j OR k=i THEN
            BT=EgnVal(k,j)
              EgnVal(k,j)=BT*COSX-EgnVal(k,i)*SINX
             EgnVal(k,i)=BT*SINX+EgnVal(k,i)*COSX
          ELSE
            AT=EgnVect(k,j)
             EgnVect(k,j)=AT*COSX-EgnVect(k,i)*SINX
            EgnVect(k,i)=AT*SINX+EgnVect(k,i)*COSX
            BT=EgnVal(k,j)
             EgnVal(k,j)=BT*COSX-EgnVal(k,i)*SINX
             EgnVal(k,i)=BT*SINX+EgnVal(k,i)*COSX
          END IF
```

```
      NEXT
        XT=2#*EgnVect(j,i)*SINX*COSX
       AT=EgnVect(j,j)
       BT=EgnVect(i,i)
         EgnVect(j,j)=AT*COSX2+BT*SINX2-XT
        EgnVect(i,i)=AT*SINX2+BT*COSX2+XT
         EgnVect(j,i)=(AT-BT)*SINX*COSX+EgnVect(j,i)*(COSX2-SINX2)
        EgnVect(i,j)=EgnVect(j,i)
      FOR k=1 TO
n:EgnVect(j,k)=EgnVect(k,j):EgnVect(i,k)=EgnVect(k,i):NEXT
       END IF
103 NEXT j
102 NEXT i
      IF DidRotate% GOTO 3    'Look for the next large off-diagonal element
     IF thr>FORM GOTO 23
      'Sort the eigenvectors and eigenvalues
     FOR i=2 TO n
       j=i
29     IF EgnVect(j-1,j-1)>=EgnVect(j,j) GOTO 110
        SWAP EgnVect(j-1,j-1),EgnVect(j,j)
       FOR k=1 TO n:SWAP EgnVal(k,j-1),EgnVal(k,j):NEXT
      j=j-1
      IF J>1 GOTO 29
110   NEXT i
END SUB
```

The program, like many computer programs, prints far too many significant figures. Values approximating zero are given as finite numbers with large negative exponents. The input-output form for the allyl system is as follows. Compare these results with those in Chapter 2.

```
Order of Matrix 3
?0
?1
?0
?1
?0
?1
?0
?1
?0
  1.4142........
-1.3937........
-1.4142........
.5000........          -0.7071........          -0.5000........
.7071........          -8.70........D-09        0.7071........
.5000........          0.707........            -0.5000........
```

References

1. Mathcad is a product of the MathSoft.Inc., 201 Broadway, Cambridge, MA 02139 (phone 1-800-Mathcad) available for both Macintosh and DOS machines. The author found the "Mathcad Treasury of Methods and Formulas" by Paul Lorczak to be a valuable aid for the novice.

2. QuickBASIC and MS DOS are trademarks of the Microsoft Corporation. The author wishes to thank Dr. Arthur Busby for adapting this program from an older subroutine written in another language.

Index